A History of Oxford Anthropology

Methodology and History in Anthropology

General Editor: David Parkin, Director of the Institute of Social and Cultural Anthropology, University of Oxford

A HISTORY OF OXFORD ANTHROPOLOGY

Edited by
Peter Rivière

Berghahn Books
New York • Oxford

First published in 2007 by

Berghahn Books

www.berghahnbooks.com

©2007, 2009 Peter Rivière
First paperback edition published in 2009

Library of Congress Cataloging-in-Publication Data

A history of Oxford anthropology / edited by Peter Riviere.
    p. cm. -- (Methodology and history in anthropology)
Includes bibliographical references and index.
ISBN 978-1-84545-348-0 (hbk) -- ISBN 978-1-84545-699-3 (pbk)
1. Anthropology--England--Oxford--History. 2. Anthropologists--England--
Oxford--History. 3. Tylor, Edward Burnett, Sir, 1832-1917. 4. Balfour,
Henry, 1863-1939. 5. Pitt-Rivers, Augustus Henry Lane-Fox, 1827-1900.
6. Radcliffe-Brown, A. R. (Alfred Reginald), 1881-1955. I. Rivière, Peter.
GN17.3.G7H57 2007
301.09425'74--dc22

                                                        2007034654

British Library Cataloguing in Publication Data
A catalogue record for this book is available from the British Library

ISBN 978-1-84545-348-0 (hbk) -- ISBN 978-1-84545-699-3 (pbk)

To all those who played a part in the success of Oxford anthropology but whose contribution we have not managed to fit into this book.

The first Diploma students in Anthropology on the occasion of their practical examination in 1908. Left to right Francis Knowles, Henry Balfour (examiner), Barbara Freire-Marreco and James Arthur Harley. In the background are some crossbow displays in the Upper Gallery of Pitt Rivers Museum. Copyright Pitt Rivers Museum, 1998.266.3.

CONTENTS

LIST OF FIGURES

LIST OF CONTRIBUTORS

Jonathan Benthall, sometime Director of the Royal Anthropological Institute.

John Davis FBA, Warden of All Souls College, Oxford. Sometime Professor of Social Anthropology, University of Kent; sometime Professor of Social Anthropology, University of Oxford.

Christopher Gosden FBA, Professor of European Archaeology, University of Oxford, Fellow of Keble College, Oxford.

Geoffrey Harrison, Professor Emeritus of Biological Anthropology, University of Oxford, Fellow Emeritus of Linacre College, Oxford.

Wendy James FBA, Professor of Social Anthropology, University of Oxford, Fellow of St. Cross College, Oxford.

Frances Larson, Researcher at Pitt Rivers Museum.

Alan Macfarlane FBA, Professor of Anthropological Science, University of Cambridge, Fellow of King's College.

David Mills, University Lecturer in Pedagogy and the Social Sciences, University of Oxford, Fellow of Kellogg College.

Robert Parkin, Temporary Lecturer in Social Anthropology, University of Oxford.

Alison Petch, Registrar and Researcher at Pitt Rivers Museum.

Peter Rivière, Professor Emeritus of Social Anthropology, University of Oxford, Fellow Emeritus of Linacre College, Oxford.

THE MAGIC OF OXFORD ANTHROPOLOGY

Alan Macfarlane

The essays in this volume have approached the history of anthropology in Oxford mainly from an administrative and chronological perspective. This is commendable, but it is also worth looking at the matter from an anthropological viewpoint. Here I shall try very briefly to ask the question of why Oxford has arguably contributed more to our understanding of tribal societies than any other department of anthropology in the world. I shall take a participant-observer position. I spent over twelve years of my life in Oxford, at school and University, where I read history. For the last thirty-five years I have been at Cambridge, observing Oxford from the outside. Of necessity I shall be very selective, in particular omitting the role of Biological Anthropology, the Pitt Rivers Museum and a number of distinguished Oxford social anthropologists.

David Hume summarised the greatest problem of anthropologists. 'Let an object be presented to a man of never so strong natural reason and abilities; if that object be entirely new to him, he will not be able, by the most accurate examination of the sensible qualities to discover any of its causes or effects' (see Winch 1971: 7). The greater the gap in experience, the more the incomprehension. The problem was also put by the Oxford philosopher R.G. Collingwood when he wrote that 'though we have no lack of data about Roman religion, our own religious experience is not of such a kind as to qualify us for reconstructing in our own minds what it meant to them' (1946: 329).

The worlds which anthropologists visited in the great 'tribal' phase were of a special kind. They were not separated into institutional spheres and they were without any particular determining infrastructure. There was no 'polity', 'economy', 'religion', 'society'. It is easy enough to acknowledge this intellectually. Yet to feel in the blood how such a world works, to get inside the system in the way Collingwood advocated is extremely difficult. How could dons living in what would seem to be the most divided and early institutionalised society on earth have any chance of understanding this in the Weberian or Collingwoodian sense of 'understanding'? England, the first large nation on earth to industrialise and the quintessence of scientific progress, with hyper-separations and extreme individualism, seems the most unlikely place from which to launch an expedition to comprehend the 'Other' of integrated, embedded, pre-Cartesian humanity. Within England, Oxford, the home of rational, highly individualistic and idiosyncratic middle-class dons, the extreme within the extreme.

Yet the reason that we can place Oxford at the forefront of the mapping of tribal worlds seems to be the peculiar nature of Oxford academic culture, and in particular the Oxford collegiate system. This provided a lived experience of 'tribal' life, of integration and non-separation.

One central feature of the tribal societies which were illuminated by Oxford anthropologists was their corporate nature. Edward Evans-Pritchard and his contemporaries and successors were able to transfer the abstract, armchair insights of the great Victorian thinkers from Morgan, Maine, and Robertson Smith and later Durkheim, to their field ethnographies because they knew what it was like to live in a working corporation. The accounts of Nuer, Dinka or other kinship systems, with their mixture of blood and fictional ties, based on a jural continuation of an entity through time, is almost identical to an Oxford college. Instead of cattle and women, the corporate property consists of buildings, lawns, libraries, wine cellars. The idea of corporate existence in fellowship, with its feeling of continuity and shared participation in a larger, co-owned whole was familiar enough. Such a lived experience is rare in the west, normally only to be found in other bounded communities such as monasteries or other 'total' institutions.

Hence some of the most insightful studies of the workings of lineages and the group nature of marriage come out of an experience of how corporate groups work. The embeddedness of social relations in a multi-functioning whole, which is simultaneously an economy, a polity, a ritual unit and an intellectual world, made it possible to grasp something as unfamiliar as African or Pacific lineage systems. It was true of both worlds that 'custom is king', and that multi-stranded interpersonal relations based on inclusion and exclusion form the infrastructure. Just as I find it easy to explain to visitors from China or Nepal how my college in Cambridge works by analogy to what I have seen in those places, so when the Oxford anthropologists travelled imaginatively into other seemingly remote worlds, in fact much was familiar.

It was the same in trying to understand tribal power. I remember Evans-Pritchard telling me that when he first laid out at the Malinowski seminar the working of the balanced oppositions and mutual tensions, which held an acephalous society like the Nuer together, Malinowski said it could not work. Indeed, looked at from the centralised, hierarchical model which the London School of Economics has represented, it was indeed impossible to feel how this would be possible. Yet the integration through factional oppositions, through fission and fusion, through feud and balanced pressures which is characteristic of societies without the state, is also the key to University and college politics in Oxford.

The Head of House is like the Leopard Skin Chief, with some authority, but little or no power. There is no police force, army, or courts of law. Everyone who has been elected is equal in the democracy of the fellowship. There is no formal hierarchy. Nevertheless the college does not tend to fall to bits, but usually coheres into an extremely effective scholarly unit through the political mechanisms which Evans-Pritchard and Fortes and others described. Rivalries, micro-alliances and the occasional use of brokers and adjudicators similar to the 'saints' in nomadic societies maintain order in an apparently almost effortless way. An Oxford college is the perfect tribe, riven internally with half-hidden tensions, yet preserved by these very differences and presenting, when faced with an outside threat, a great deal of unity.

The most difficult of all the tasks facing western anthropologists, and the area where perhaps Oxford anthropologists have made their greatest contribution, is in the understanding of tribal beliefs. When faced with polytheistic, shamanistic, enchanted and magical worlds, how could the products of two thousand years of increasing separation and rationalisation identify with ideas which had been long over-laid with a mixture of Greek, Christian and scientific rationality? Yet Oxford anthropologists are famous for their pioneering studies of myth, religion, magic and witchcraft in the series of works from Marett through Evans-Pritchard down to the present.

That Oxford is itself 'magical' is obvious, yet difficult to document. It is something I myself experienced as a pupil at preparatory school in North Oxford, then as an undergraduate and graduate in Oxford. As I wandered the city and surrounding countryside with Matthew Arnold's 'Scholar Gipsy' or Lewis Carroll's 'Alice' book in my hand, I felt the magical lands just on the other side of modernity. As I discovered the 'Oxford' magical school of Dorothy Sayers, C.S. Lewis and J.R.R. Tolkein and their parallel worlds, I did not find it impossible to preserve beside the western rational divisions a different enchantment. More recently Philip Pullman and others have set their work in Oxford for similar reasons and Harry Potter's Hogwarts was filmed in a great Oxford college hall. It is perhaps no coincidence that R.G. Collingwood's posthumous collection of essays should be called *The Philosophy of Enchantment*.

So Oxford takes magic seriously, and when anthropologists went to lands where magic and witchcraft are so important they did not dismiss it as irrational nonsense, but spread it out so we can marvel at other worlds and other interconnections.

Likewise, Oxford is a very 'religious' place. Religions of many kinds, with their accompanying myths and rituals aplenty, have been preserved in a way which is unusual. Oxford is famous both for its high church legacy, as well as its evangelical Christianity, tolerant and almost polytheistic, yet devout and serious. It provided a perfect home for many of the great anthropologists, almost all of them Catholics or Jews, as Evans-Pritchard pointed out to me, who have so widened our understanding of religions in tribal societies; Tylor, Steiner, Evans-Pritchard, Godfrey Lienhardt, Meyer Fortes, Max Gluckman, David Pocock and Mary Douglas are just a few of the Oxford-associated figures.

While my own university in Cambridge, with its more secular, scientific and rationalist flavour has contributed greatly to the study of peasant societies, to power, economics, to literacy and other fields, Oxford and its pupils have done more to make us understand the central role of symbols, rituals, mythologies in embedded tribal worlds than anywhere else I can think of.

A small group of a few dozen teachers and their students over a period of a hundred years has opened up worlds, the huge diversity and richness of systems which would otherwise have disappeared in patronising neglect. In the great tradition from Montaigne and Montesquieu and the Enlightenment thinkers, a few deep thinkers have been able to transcend the vast distance between a world of wine, libraries, bells and academic discussion, to places where humans seemed so very different. They showed the psychic unity of mankind and treated their informants as friends, as colleagues, as fellows as it were, and thereby expanded our horizons. It is now too late to do this in most of the world. If Oxford had not existed, we would have had to invent her, for in the world of anthropology her scholars gave us as rich imaginative alternatives and as plausible systems of operation as anything encountered by Alice through the looking glass.

Finally, it is worth noticing that despite the uncertainties and struggles so well documented in the essays in this book, Oxford gave many the security to think grandly and with creative originality. I remember long ago discussing the reasons why King's College in Cambridge was the most radical college in that University in the 1970s. It was suggested to me that with its magnificent Chapel and ancient history it did not need to prove itself. It could be adventurous because it was so secure in its superiority. The same could be said of Oxford as a whole. Famous for its aristocratic idiosyncrasies captured for many in *Brideshead Revisited*, with its political connections through numerous Prime Ministers and other powerful establishment figures, it is the very emblem of upper middle class power. An Oxford man (or woman) did not have to prove themselves.

So when Oxford-trained or connected fieldworkers went to live in societies at the other extreme of material affluence and political power, or settled down to write about apparently absurd and counter-intuitive systems of thought, they did not suffer from too much self-reflexive lack of self-confidence. The tone of self-assured good sense that amused (or irritated) Clifford Geertz in his critique of Evans-Pritchard is the very thing that made it possible to describe several impossible things before breakfast, or at least at dinner, and not to fear incredulous and destructive laughter.

Albert Einstein once remarked that 'Unless an idea starts off as absurd, there is no hope for it'. Through creating a virtual community, by uniting their work and their lives, by their assurance, generations of Oxford scholars have been able to make the absurd leaps which take us into new and previously unsuspected worlds. They had the privileges, the shared zeal and the shock of similarity-with-difference which engenders true creativity and they made good use of it. It has been an honour to be on the fringes of such a group for over forty years. Our students are still thrilled to read the works of these great masters of the discipline.

INTRODUCTION

Peter Rivière

The formal recognition of anthropology's existence at Oxford University occurred in the spring of 1905 with the promulgation of a statute creating 'a Committee for the organization of the advanced study of Anthropology, and to establish Diplomas in Anthropology to be granted after examination' (*Oxford University Gazette* (hereafter *OUG*) 1904–5: 536). The Committee for Anthropology, as it was known, met for the first time on 27 October 1905. The year 2005 thus seemed an appropriate and suitable occasion on which to celebrate the centenary of the subject at the University. This volume is composed of the contributions made to a workshop on the history of anthropology at Oxford University which was held in conjunction with the centenary celebrations.

For some years I had been planning to write a history of Oxford anthropology with a view to its publication around the date of the centenary. In the event, another major publishing commitment made it clear to me that, as a single-handed project, this was unrealisable. On the other hand, as the plans for a centenary conference developed, a session devoted to the history of Oxford anthropology was included. This volume contains the proceedings of that session held on 16 September 2005. Thus my original intention became a reality on the back of shared labour. I have no doubt that the various contributors, with their different approaches and insights, have brought to the volume a far broader, more many-sided picture than I alone would have achieved.

A criticism might be made, and, knowing academics, almost certainly will be, of the choice of contributors; that it is too much an 'insiders' history. Indeed, with two exceptions, everyone of the contributors is or has been closely associated with Oxford anthropology.[1] To a large extent this was quite deliberate as I had been aware that those whom I approached had an interest in the topic and, in some cases, had been actively researching the specific period or individual. Basically, a chronological framework has been adopted, and it has to be admitted that the emphasis is on social anthropology, although this becomes more marked following Radcliffe-Brown's (hereafter R-B) attempts to distance social anthropology from ethnology, physical anthropology and linguistics. Before that it is much easier to treat anthropology in the round as it involved all aspects of the subject.[2] To compensate for this emphasis, there is a chapter devoted to physical, later biological anthropology, and another to the Oxford University Anthropological Society and the *Journal of the Anthropological Society of Oxford* (*JASO*), which, rather confusingly, are unconnected.

Other than agreeing the period or topic, contributors did not receive any guidelines about how they were to tackle it or what to cover. The result has been a variety of approaches with contributors rightly concentrating on those aspects from their period which they find most significant. This has resulted in a history where the reader may well be left wondering what is going on backstage while the events on stage, those covered in the chapters, unroll. Accordingly, in this Introduction, I will try to fill out the wider picture. This approach has the advantage, at least for me, of my being able to offer some of the personal reflections that I would have included had I written my own history of anthropology at Oxford.

There does appear to be one striking omission from this volume, the Pitt Rivers Museum. This is quite deliberate for the simple reason that the Museum is the senior partner by twenty-one years and celebrated its own centenary in 1984.[3] Thus, while 1905 may be identified as the formal birth date of anthropology, as with the birth of everything, there was a period of gestation; in the case of anthropology it was particularly long and difficult. As Christopher Gosden, Frances Larson and Alison Petch show in Chapter One, 'Origins and Survivals: Tylor, Balfour and the Pitt Rivers Museum: their Role within Anthropology in Oxford 1883–1905', the history of anthropology at Oxford University started many years before 1905.

An interest in anthropology in the University can certainly be traced back to the 1860s. This was a period of intense debate over geological time and biological evolutionism. Trautmann (1992) has argued that there was a simultaneous revolution in ethnological time; certainly the decade was particularly fruitful for anthropology. Maine's *Ancient Law* was published in 1861, and 1865 saw the appearance of no fewer than three seminal works in the development of anthropology. They were Lubbock's *Prehistoric Times*, McLennan's *Primitive Marriage* and Tylor's *Early History of Man*. The founding in Oxford in 1867 of an anthropological society may well have been a response to this increased interest in the subject.[4] How long the society survived is unknown, but there were at least two senior members of the University who had an interest in anthropology and were keen to promote it. One of these was George Rolleston, Linacre Professor of Anatomy and Physiology, and the other Henry Moseley, who after Rolleston's death, became the Linacre Professor of Human and Comparative Anatomy. Both men were members of both the Anthropological Institute and the Ethnological Society in London. Rolleston, in particular, was influential in persuading Pitt-Rivers to offer his collections to Oxford and in urging the University to accept them. It would seem that those who supported the transfer of Pitt-Rivers' collections to Oxford saw this as a means of forwarding their aim to introduce anthropology on to the syllabus.[5] Thus, in 1881, Moseley remarked in a letter to Augustus Franks, Keeper of Ethnography (among other things) at the British Museum, that the University's acceptance of Pitt-Rivers' collection 'would be of extreme value to students of anthropology in which subject we hope all men to take degrees very shortly' (PRM, Foundation & Early History MSS: Letter 5). This was overly optimistic but at least in 1885, the year after the founding of the Pitt Rivers Museum, anthropology was made available as a Supplementary Subject in the Natural Science Final Honour School (hereafter FHS).[6]

In fact, instruction in anthropology had been available to both members and non-members of the University from the previous year, 1884, when Tylor was appointed to a Readership in Anthropology.[7] Although Tylor was Reader, then Professor in Anthropology, as Gosden *et al.* point out, he was never the Curator of the Pitt Rivers Museum. Installation and care of the collections were invested in the Linacre Professor of Anatomy, Moseley, to whom, rather than Tylor, fell the responsibility of moving the collections from London. In practice, most of the work was undertaken first by Baldwin Spencer, Moseley's demonstrator, and when, in 1886, he moved to a chair in zoology in Melbourne, by Henry Balfour. In 1887 Acland and Moseley combined to get Balfour created temporary Assistant Curator and in 1891 he was made Curator of the Pitt Rivers Museum.

The next serious attempt to promote the interests of anthropology occurred in 1895 when Tylor and his supporters petitioned for anthropology to become a full FHS. This proposal was rejected, but such attempts are a recurring theme throughout this volume. Virtually every contributor records a move to establish an undergraduate degree in anthropology, something that was not achieved until its involvement in Human Sciences in 1970 and Archaeology and Anthropology in 1992.[8] David Mills, in Chapter 4, 'A Major Disaster to Anthropology? Oxford and Alfred Reginald Radcliffe-Brown', suggests that this was not an entirely bad thing, and that Oxford's relative dominance in the decades following the Second World War was partly owing to the lack of the distraction of undergraduate teaching.

The difficulties that anthropology faced in getting accepted clearly reflected doubts in many minds about the nature of the subject. If seen as the study of humankind in its broadest sense it was too wide and threatened the territories of some already established subjects. On the other hand, if the boundaries of the discipline were drawn too narrowly and it was limited to the study of past and present 'primitive' people, then it was barely an appropriate subject for an undergraduate degree. There was, however, more to it than this, for anthropology fell victim to an essential debate within the University, which in one form or another has still to be resolved. Gosden *et al.* refer to it as a tension between the humanities and the sciences, but it was also and remains a tension between the University and its departments on the one hand and the colleges on the other. Today this tension takes many forms, from disagreements when making appointments between departments needing specialist researchers and colleges requiring generalist teachers to disputes over fund-raising and who has the right to approach possible benefactors.

The place of anthropology within the universities had been greatly enhanced in the late 1890s by the Cambridge University's Torres Straits expedition. Furthermore the scheme in 1905 was to introduce anthropology as a graduate qualification, a marginal activity at the time, and to deal solely with 'primitives'; these compromises meant that no vested interests were threatened. Accordingly the Committee for Anthropology came into being with little fuss, and once it had completed its first task, to design the Diploma course, there were thirty years of remarkable stability. This period is dealt with by Peter Rivière in Chapter 2, 'The Formative Years: the Committee for Anthropology 1905–38'. Whereas, at the introduction of the subject, lectures and instruction were listed as being available from a wide range of people, including the professors of jurisprudence, Sanskrit, philology, Celtic and Russian, this gradually declined and the subject came to be dominated by a triumvirate. They were Arthur Thomson, Dr Lee's Professor of Human Anatomy who, until his death in 1935, was responsible for physical anthropology; Henry Balfour, Curator of the Pitt Rivers Museum until his death in 1939, who looked after the archaeological, ethnological and technological aspects of the Diploma; and Robert Ranulph Marett, Reader in Social Anthropology and Rector of Exeter College, who covered the sociological side.

To obtain the Diploma candidates had to satisfy the examiners in anthropology writ large, that is to say physical anthropology, ethnology and archaeology and social anthropology, although a

Certificate was available in any one of the specialisations. Thus while the subject had a disciplinary unity it was spatially dispersed. Physical Anthropology was located in the Department of Human Anatomy, where, from 1927 it became known as the Laboratory of Physical Anthropology, with Leonard Dudley Buxton as Reader in the subject. The Department of Ethnology,[9] which came into being in the nineteenth century as a sort of adjunct of the Pitt Rivers Museum, was located with it in the University Museum. Social Anthropology had a much more nomadic existence. To begin with, because of the few involved, space was found in Exeter College, but the numbers outgrew that and in 1914, the year in which permission to use the name Department of Social Anthropology[10] was given, its home became Barnett House, 26 Broad Street. From there, in 1920, it joined with the Geographers in Acland House, 40 The Broad (see Figure 3), and in 1936, when Acland House was to be demolished to make way for the New Bodleian, it followed the Geographers to 1 Jowett Walk, on the corner of Mansfield Road, where, until this year (2005), the School of Geography has remained.

The interwar years saw stagnation, if not decline, in anthropology both at Oxford and Cambridge. The active centre of anthropology had moved to London, where, to oversimplify, theoretical positions divided the functionalist London School of Economics, ruled over by Bronislaw Malinowski, from the diffusionist University College London, home to Grafton Elliot Smith and William Perry. If the former proved more successful, this was, as Goody has pointed out (1995: Chapter 1), because of his access to financial resources, specifically his close relationship with the Rockefeller Foundation. It was through this Foundation's support for the International African Institute that Malinowski was able to attract through research funding many of those who rose to the top of the subject after the Second World War. Oxford anthropology basically missed out on the largesse available from the Rockefeller Foundation except in one very important respect; it was Rockefeller which funded the research lecturership that Evans-Pritchard (hereafter E-P) took up in 1935.

It is the case that by the 1930s the triumvirate at the heart of Oxford anthropology was growing old and the question of the succession loomed. Nor was anthropology's future secure, as with the possible exception of Balfour, teaching anthropology was neither Thomson's nor Marett's day job. Thomson died in 1933 and Balfour in 1939, the former to be succeeded by Wilfred Le Gros Clark and the latter by Thomas Penniman. The other posts in anthropology were those of Buxton as Reader in Physical Anthropology and Beatrice Blackwood as Demonstrator in Ethnology. There was, however, no statutory post in social anthropology for Marett's official position was held at Exeter, and after a certain amount of toing and froing the University agreed to create a statutory readership to replace Marett. It was here that All Souls College stepped in and offered to put up the money to convert this into a chair. How this came about is the topic of John Davis's contribution, 'How All Souls got its Anthropologist' (Chapter 3). To understand this Davis has had to move well outside the sphere of Oxford and anthropology and the cast of characters he introduces are figures on the much wider stage of empire. All Souls' interest in anthropology was directly related to colonial and imperial administration, and the college's links led to Whitehall, Westminster and beyond.

This is possibly a convenient point at which to consider a little more closely the relationship between Oxford anthropology and colonial rule, as this is a subject that crops up in several of the chapters. As Rivière shows, one of the arguments advanced for the introduction of the Diploma in Anthropology at the very beginning of the twentieth century was its value to overseas administration. He also refers to the frequent attempts between 1896 and 1921 to set up national centres of applied anthropology whose purpose was to help with the government of empire. These efforts, in which Oxford was directly involved, got nowhere and the provision of courses for overseas administrators was left to the individual university departments. From 1908 onwards these administrators formed an important component of the anthropology graduates at Oxford. That the right to teach them was a valuable asset is made quite clear by Davis who also shows how the Royal Anthropological Institute (hereafter RAI) tried to muscle in on the business on the side of London University. Indeed, it is possible to see the creation of the Chair in Social Anthropology at Oxford as associated both with the attempt by London University to monopolise the teaching of colonial cadets and with the failure to found an African Institute at Oxford.

Nor after R-B took up his chair at Oxford in 1937 did this matter go away. As Mills documents, both in Oxford, in his dealings with Nuffield College, and as President of the RAI, he found himself concerned with anthropology's interest in colonial affairs. A bid to the University Grants Committee in 1944 for increased resources for anthropology was argued on the basis of a need for such expertise in the colonies following the war. The teaching of colonial cadets continued after the Second World War, but, with the disappearance of the Empire and thus the need for overseas administrators, the course was terminated in 1962. Remarks on some of the consequences of this will be taken up later in this Introduction.

The arrival of R-B as the first incumbent of the new Chair in Social Anthropology seriously disturbed the comfortable routine into which the subject had lapsed, as Mills describes in Chapter 4. Even before his arrival, he engaged in a discussion with the Committee for Anthropology on the introduction of a FHS. Unfortunately their views on the nature of this degree did not coincide. The members of the Committee devised the FHS with a similar academic content to the Diploma, a general anthropology. R-B, however, wanted to reform anthropology in line with his own views on the subject; he wanted the degree to specialise in social anthropology. At the same time he wanted to change the format of the Diploma in a not dissimilar way in order to allow increased specialisation. Unable to change the view of the Committee for Anthropology on the FHS, he concentrated on reforming the Diploma. What he proposed was four separate Diplomas, one each in social anthropology, human biology, prehistory, and comparative technology. This scheme was put forward and rejected in 1938–9, as was an almost identical proposal the following year, on the grounds that effort should rather be put into devising a FHS in Anthropology. In 1940 – and it is amazing that anyone was worrying about the nature of a diploma in anthropology that year – a new format was agreed, whereby all candidates would sit two papers and a practical examination on general anthropology and then further papers and a thesis in one of three specialisations, physical anthropology, or prehistoric archaeology and comparative technology, or social anthropology (*Examination Statutes*, 1940: 374). The outbreak of war and R-B's absence in São Paulo prevented any further changes that he might have had in mind. He returned to Oxford in late 1944, by which time he was too close to retirement in 1946 to have any further impact.

While R-B was away, Daryll Forde, then Professor at Aberystwyth but seconded to the Foreign Office Research Department in Oxford, looked after the Institute. Among other things he prepared in 1944 the bid, already mentioned, to the University Grants Committee. He argued that because of changing colonial conditions anthropology would have an even more important part to play after the war. He therefore proposed an establishment of a professor, a reader, one senior lecturer, one junior lecturer, a secretary/librarian, and a £2500 research fund. Although the staff did increase later in the decade, in the last year of R-B's tenure, the Institute was still very small, consisting of the professor, one lecturer, a secretary – librarian, and seven students.

Evans-Pritchard succeeded R-B in 1946. It is the former's years as professor in Oxford which form the topic of Wendy James's chapter, 'A feeling for form and pattern, and a touch of genius'. This was the period that is so often regarded as the Classical or Golden Age of Oxford social anthropology, but James argues that the myth-makers have been at work to obscure some of the realities. For example, she argues that the Institute was not so much a place of fermenting new ideas, but one where the full implications of E-P's earlier works received proper recognition and application. Furthermore, even if Oxford anthropology had achieved a central role in the subject both in Britain and abroad, it was still seen as marginal within the university. E-P's attempt to introduce a FHS in the subject was dismissively snubbed and Sociology, which would not get its own chair for many more years, turned its greedy eyes, in the name of relevance, on that of Social Anthropology. There is no doubt, however, that by the time E-P retired in 1970, the Institute was on a secure foundation, many times the size, in both staff and students, that it had been in 1946.

It had in this period found a relatively permanent home. When E-P took over, the social anthropologists were still housed with the geographers at the Mansfield Road/Jowett Walk site, but increasing demands for space by the latter soon saw the former on the move again. In 1948, Social Anthropology moved to Museum House on South Parks Road which had been Tylor's home (see Figure 4). These premises were rather more spacious and provided room for a library, periodicals room, lecture room, and rooms for the secretary/librarian and five teaching staff. Once again, however, Social Anthropology was only one step ahead of the demolition crew, and in 1952 it moved to 11 Keble Road to make way for Inorganic Chemistry. This proved no more permanent, for by the 1960s there were plans to demolish all the houses on the north side of Keble Road for an extension of Physical Sciences. These houses were saved and a vastly extended 11 Keble Road is now the Computing Laboratory with a Parks Road address. However, the decision not too demolish the houses was only made after the Institute had moved, which it did in the winter of 1965–6, to 51–3 Banbury Road.[11]

During E-P's reign, there was a steady turnover, and increase, in staff. Fortes came and went to Cambridge; and Gluckman came and went to Manchester. Other people who held posts for longer or shorter periods included Franz Steiner, M.N. Srinivas, Louis Dumont, David Pocock, Mary Douglas (then Tew), Paul Bohannan and John Peristiany. Those who had arrived and were still there when E-P retired were John Beattie, Godfrey and Peter Lienhardt, Rodney Needham, Edwin Ardener and Ravi Jain, with John Campbell as a Fellow of St Antony's College.

As already mentioned, the course for overseas service cadets was discontinued in 1962. As James comments, the expectation was that this would result in a decline in anthropology's fortunes as arguments in support of the subject's practical value weakened. In fact, the opposite happened and the following two decades, approximately until 1980, were a boom time for anthropology across the United Kingdom.

It was a period, as I remember it, when there was little discussion about the practical relevance of the subject and rarely was it called upon to justify itself in this way. Existing departments expanded and new departments were created.[12] Research money was relatively easy to come by and there was a plentiful supply of graduate studentships available from the Social Science Research Council (SSRC) following its formation in 1965. Oxford, as the only department in the country which concentrated solely on graduate students, did very well out of this. In 1970, it received thirteen new SSRC studentships and between 1975–80 it produced forty-three doctorates, compared with Cambridge's thirty-three, and as many as the LSE, SOAS and UCL combined (see Webber 1983). This situation, however, was already radically changing by 1980 with an ever growing number of departments opening graduate courses and demanding studentships, and thus threatening Oxford's relative dominance in the production of professional anthropologists.

During this period, as James describes, there was a further unsuccessful attempt to introduce a FHS in Anthropology. The proposal envisaged an anthropology writ large, and its failure seems to have resulted in the abandonment of this aim[13] and a return towards the reform of the Diploma that Radcliffe-Brown had tried to introduce before the Second World War, with separate diplomas for each anthropological specialisation. Although this was not fully achieved during E-P's tenure of the chair, between 1962 and 1974 there were numerous attempts to modify the format of the Diploma, all of them unsatisfactory. The details can be safely left to one side, but the important changes were the following: in 1964, the single Diploma in Anthropology was replaced by four diplomas, in Human Biology, Prehistoric Archaeology, Ethnology, and Social Anthropology but they all retained one paper in common.[14] At the same time the Certificate was scrapped, which had anyhow become little more than a consolation prize (*Examination Statutes*, 1964: 526–30). After various other modifications, finally, in 1974, four completely independent diplomas came into existence. This struggle for separate diplomas was to a large extent driven by the social anthropologists, but how far this still reflected a pursuit of R-B's views and how far it was part of a numbers game, it is difficult to ascertain. Certainly, in the latter, social anthropology came out much better than the other subjects. For example, in 1966, there were thirty-four people reading for the Diploma in Social Anthropology, two for that in Human Biology, two for Ethnology and one for Prehistoric Archaeology. This pattern was fairly typical of the decade.

After James's chapter, the volume veers away from its chronological course and takes up two topics that are important in the history of

Oxford anthropology. The first is Geoffrey Harrison's 'Oxford and Biological Anthropology'. Physical or Biological Anthropology made its formal appearance at Oxford in 1885 when Anthropology became a Supplementary Subject in the FHS of Natural Science. Although the subject included components on archaeology, philology and cultural development it was predominantly physical anthropology. Furthermore it was two successive professors of human anatomy who happened to have an interest in physical anthropology that kept the subject alive as a part of the Department of Human Anatomy. It took a tentative step towards an independent existence in 1927 when Leonard Dudley Buxton was appointed to a Readership in Physical Anthropology and the Laboratory of Physical Anthropology, still under the auspices of Human Anatomy, was founded. Full autonomy was delayed until 1976 when the Department of Biological Anthropology, headed by Harrison himself, was created and moved to 58 Banbury Road.[15] In 1990 the Department changed its name to the Institute of Biological Anthropology (IBA) as part of a re-organisation that saw the creation of a School of Anthropology and Museum Ethnography, consisting of IBA, the Institute of Social and Cultural Anthropology (ISCA) formed from the Department of Ethnology and Prehistory and the Institute of Social Anthropology, and the Pitt Rivers Museum.

IBA's autonomy has, however, proved fleeting. On Harrison's retirement in 1994, the new professor, Ryk Ward, a geneticist, had an entirely different academic agenda and increasingly allied the Institute of Biological Anthropology with the Department of Zoology. Ward's sudden and premature death in 2003 and a series of moves by other staff have left Oxford Biological Anthropology virtually moribund. The rump of the department is now physically and administratively under the wing of Zoology and still awaiting the appointment of a new professor. As Harrison points out, it is not clear what the future holds for Biological Anthropology, and he sees the current syllabus as a similar if modernised version of that for the Diploma in 1905. On the other hand, the association of the Lecturers in Human Ecology (Stanley Ulijaszek), originally located in IBA, and in Medical Anthropology (Elizabeth Hsu) with ISCA has done much to advance the return to a holistic anthropology that was another theme of the centenary conference (see Parkin and Ulijaszek, 2007).

Robert Parkin's account of the Oxford University Anthropological Society (OUAS) and the *Journal of the Anthropological Society of Oxford* (*JASO*) covers the history of two other aspects of the vitality of Oxford anthropology. The existence of an anthropological society in Oxford in the 1860s has already been mentioned but it seems to have been short-lived, and the present OUAS dates from 1909. JASO – and the

'Society' of its title, as Parkin explains, has only a tangential connection with that in OUAS – first appeared in 1970. Although *JASO* has often suffered from slippage in its publication date, it survived until the centenary year as one of the longest running student anthropological journals. Whereas, from its outset, *JASO* was very much a student affair, the OUAS was for much of its life dominated by senior members of the university and its membership spread well beyond the limited numbers reading or teaching anthropology. After a period of hibernation at the end of the twentieth century, in its present, revived form, the OUAS is very much a junior members' initiative for junior members.

This raises another aspect where the history of Oxford anthropology reflects changes in anthropology more widely, namely the degree of specialisation that has occurred in the subject. The professionalisation of the subject has been plotted on the national level by Kuklick (1991) who sees it as but one manifestation of a changing occupational structure. Gosden *et al.* subtitle one section of their paper 'Participatory anthropology', in which they indicate just how much the collections of the Pitt Rivers Museum derive from the activities of people who were not only not anthropologists, but not academics at all. Alongside this professionalisation, there had been increasing specialisation. For social anthropology this has its roots in Malinowski's LSE, and we have seen how it was taken up by R-B in his attempt to introduce a specialised social anthropology qualification at both graduate and undergraduate level. R-B's presidency of the RAI, as Mills describes, saw an internal dispute over an attempt by social anthropology to obtain a monopoly over the teaching of colonial cadets. This failed, but after the war social anthropologists, led by E-P, declared at least semi-autonomy by founding their own professional body, the Association of Social Anthropologists (ASA). The ASA had strict criteria of eligibility for membership, and its annual vetting of new members regularly defined or re-defined the nature of the subject.[16]

A further symptom of this move to increasing specialisation is also to be found in a decision by the ESRC (as the SSRC was re-named in 1982) to concentrate on providing further training for those with anthropological qualifications and not to fund conversion courses. The list of anthropologists whose first degrees were in a different subject and who then converted to anthropology is long and illustrious. The Diploma, and its successor the Master of Studies (M.St.), was quite specifically a conversion course and during the second half of the twentieth century drew in people with academic qualifications of every sort and none.[17] James refers to the increasing interaction with history, philosophy and theology during the years of E-P's tenure, and this is further evinced by the number of Diploma students with just those backgrounds. That this was good for the health and development of the subject, there can be no doubt. The number of studentships provided by the ESRC is now so few that this decision in itself may not have an impact on the number of enrolments, but it has closed the door to a source of excellent future anthropologists and narrowed the field.

The Oxford University Anthropological Society appears to have followed a similar trajectory. As Parkin notes, it was in the beginning a genuinely university-wide society and remained so until after the Second World War. Its officers and those who were members and attended its meetings where drawn from a cross-section of interested people. It was only later that it became increasingly a society for senior and junior members of the university who were also anthropologists. In its most recent manifestation, a society mainly for anthropology graduates, it seems on occasion to operate as a kind of 'alternative institute'. Although, perhaps this is too recent for it to be possible to have a proper perspective on it, this may reflect an increasing distance between staff and students within ISCA as a result of the growing pressures, often externally derived but internally exerted, which allow less and less time for the socialising that was such a feature of the Institute, not simply during the post-Second World War decades, but even from the earliest years, as Wallis (1957), quoted by Rivière, indicates.

Finally, to return to the Pitt Rivers Museum on this topic, although the work has not been done, the members of the Relational Museum team are of the impression that since 1945, though there are still a large number of donors from a wide variety of backgrounds, there are probably fewer than there were prior to 1945. Their guess is that collecting has become a more specialist activity.

The final period, the thirty-five years since the retirement of E-P, presents particular difficulties. During this time there have been four different professors and between 1977 and 1990 the Institute was overseen by a Management Committee, which was a committee of the Anthropology and Geography faculty board. It was composed of the permanent members of the ISCA academic staff, one of whom was elected as chairman and was responsible for the day-to-day running of the Institute,[18] together with the chairman of the board and one other member. As well as this there has been a number of administrative re-organisations which have continued into Oxford anthropology's second century. It was clear to me that it was essential to get an outsider to cover this period as most of the insiders had lived through those years and were participants in its events. Nor was it that easy to identify a suitable and willing outsider[19] when I was struck by the lucky idea of approaching Jonathan Benthall. As Director of the RAI for most of the period, without any particular departmental loyalty and a frequent visitor to Oxford, Benthall was well placed to see those years with a relatively dispassionate eye. In his chapter, entitled 'Since 1970: through Schismogenesis to a New Testament', he provides a clever and diplomatic survey of the past three decades.

By the time E-P retired in 1970, Oxford anthropology, particularly social anthropology, had grown enormously. By that date there were six posts including the Chair and over one hundred students. Ethnology had shown a similar increase with the creation of two Assistant Curatorships, later to be combined with University Lecturerships in Ethnology and then, when the curator was retitled Director, the posts followed suit and became Curatorships. Physical Anthropology's expansion, as described by Harrison in Chapter 6, followed in the 1970s; first, in 1972, with the creation of a second lecturership, and then, in 1976, when the Department, renamed Biological Anthropology, finally attained its independence from Human Anatomy.

Maurice Freedman took up the chair in 1970 and Beattie departed for Leiden. Rivière was appointed to the latter's post in 1971 and the following year Wendy James obtained a newly created lecturership. Jain returned to India in 1975 and was replaced by Nick Allen. With Freedman's sudden and premature death in 1975, Needham was elected to the chair the following year and his lecturer's post was filled by Robert Barnes in 1977. The next ten years saw no change in staff, and then in 1986, Peter Lienhardt died and his post, a faculty lecturership, was abolished. The following year Ardener died suddenly and unexpectedly and Marcus Banks was appointed at short notice to a temporary lecturership to fill his place. Banks then stayed on as departmental demonstrator and was finally confirmed in a tenured lecturership in 1995. Godfrey Lienhardt retired at the end of the 1987–8 academic year and Paul Dresch arrived in 1989. John Davis replaced Needham in 1990, and on his appointment as Warden of All Souls in 1995 was succeeded by David Parkin the following year. In the final years of the Oxford anthropology's first centenary, Allen took early retirement and Rivière reached retirement age – the former to be replaced by David Gellner and the latter by Elizabeth Ewart. As already mentioned, new posts have also been created – a lecturer in medical anthropology (Elisabeth Hsu) and another in human ecology (Stanley Ulijaszek) – and as the centenary year closes a second Chair in Social Anthropology has been advertised and a new University Lecturership in Social and Cultural Anthropology created.[20]

Prior to 1990, the date when ISCA came into being, the ethnology curators/lecturers from the Department of Ethnology and Prehistory had been Audrey Butt (1955–83), Ken Burridge (1959–68), and Peter Gathercole (1968–71). After1990 these posts, were, for teaching purposes, treated as part of ISCA. At that time, the ethnologists[21] were Schuyler Jones, who had followed Brian Cranstone as Director of the Pitt Rivers Museum in 1985,[22] and Howard Morphy and Donald Tayler. Jones was succeeded by Michael O'Hanlon during the academic year 1997–8, and Morphy and Tayler by Clare Harris and Laura Peers at the beginning of the next academic year.

If we widen the perspective we find a number of anthropologists appointed in other departments with specific area interests, such as Michael Gilsenan (Middle East), Roger Goodman (Japan), Frank Pieke (China) and Charles Ramble (Tibet), who maintained or maintain close ties with the School, as has the holder of the post in Development Anthropology at Queen Elizabeth House, formerly David Sneath and then Laura Rival. Also associated with Queen Elizabeth House are the Centre for the Cross-cultural Research on Women (now International Gender Studies) and the Refugee Studies Centre, both of which have always had a strong anthropological presence. A more recent and important development has been the founding, with a substantial grant from the ESRC, of the Centre on Migration, Policy and Society (COMPAS) with Steven Vertovec as its Director. It forms part of ISCA and occupies 58 Banbury Road, the former home of the Institute of Biological Anthropology. In the first year of anthropology's second centenary, this roll call was added to by a West African post to be shared with Area Studies (David Pratten) and by a Lecturer in Migration Studies to join the staff of COMPAS (Katherine Charsley).

The relatively affluent period of the 1950s, 1960s and 1970s came to an abrupt halt in the final years of that last decade. A financial squeeze was placed on the whole university system and expansion turned into contraction. The social sciences were particularly hard hit, although social anthropology weathered the storm remarkably well. It was by the standards of other social sciences, such as economics and sociology, very small, did not cost much money and had the reputation for harmless if slightly eccentric scholarship. The subject did not go entirely unscathed, although it was mainly in joint departments of sociology and social anthropology that it suffered worst.[23] Another problem was that it brought an end to the ready supply of jobs in new and expanding departments. There was much soul-searching over what sort of employment there would be for the relatively large number of doctoral students coming on to the labour market. If there had been little discussion in the previous decades about the practical use of social anthropology the topic now returned with a vengeance. With the harsher climate within the universities and funding bodies, the search for relevance became important again. Despite the gloomy outlook, the winter did not last that long in Oxford. The cuts the department suffered were not severe and the 1980s was a period of stability rather than contraction.[24]

During the past quarter-century there have been considerable changes, mainly as result of external pressures, to the degrees on offer at the Institute. The Diplomas finally disappeared in the academic year 1982–3, to be replaced by a M.St., both in Social Anthropology and in Material Anthropology and Museum Ethnography. This was merely a change in name, not in format, and was made in response to the re-titling of similar qualifications in other universities because, not surprisingly, a 'Masters' degree sounded much better than 'Diploma' to would-be students. The M.St.'s lasted until 1998 when they gave way to M.Sc.'s, which were twelve-month rather than nine-month courses and included the submission of a dissertation. At the same time a Diploma was re-introduced for those who wished to do a nine-month course, sit the written papers but not prepare the dissertation.[25] This change was made in response to outside pressures, this time from the ESRC which declined to recognise a nine-month course as providing adequate research training.

In 1981 a two-year M.Phil. was introduced as an alternative route to the D.Phil. from that of Diploma plus M.Litt. (re-titled from B.Litt. in 1979) which had been the conventional and more leisurely path.[26] This was also done in response to external factors, in this case the enormous increase in graduate fees, especially for overseas students, that was put in place in the late 1970s. The advantages of the M.Phil. are that fees paid for it count towards those for the D.Phil., and the 30,000-word thesis that it involves can be incorporated into the doctoral thesis. In recent years, the number of different taught Masters, both M.Sc.'s and M.Phil.'s, has proliferated and are now available in specialisations such as Medical or Visual Anthropology and Migration Studies.

It has been noted that many of these changes were driven by outside forces; competition for graduate students, increases in fees and changes in fee structure, and to requirements of the ESRC. In 1982 the SSRC had its name changed to ESRC.[27] This change involved the abolition of the subject committees, of which social anthropology had been one, which formed a vital link between the departments and the bureaucrats. The result has been a bureaucratisation which has impinged badly on anthropology because, as with most bureaucratic cultures, there is little appreciation of variation and the same size is expected to fit all. Thus the ESRC has interfered inappropriately in both anthropological training and research with the imposition of unnecessary requirements and penalties for non-fulfilment. The ESRC itself has also seen a decline in its funds and this is most visible in the drop in the number of studentships available for British graduates. Within the past quarter-century the fall in the number of studentships it has made available for social anthropology has seriously affected the nationality profile of most British departments. In recent years, British students at ISCA have become a minority, and those from the USA in a distinct majority, though there has been some increase from EU countries.

Another major change has been one from a majority of men to one of women students, a shift which has been matched in recent years among the staff with more women than men being appointed to full-time university lecturerships within the School of Anthropology.

Mention has been made of the various attempts to introduce a FHS in Anthropology, but after the rebuff in 1948–9, the enthusiasm for such a degree waned. Even so anthropology found itself finally involved with undergraduates via another route, through its participation in Human Sciences, which FHS was introduced in 1970 and in which Biological Anthropology in particular has played a central role. The involvement with undergraduates was taken a step further in 1992 with the introduction of the FHS in Archaeology and Anthropology.[28] The picture is not, however, entirely rosy. There is a looming threat to Human Sciences, and with much talk of a switch in emphasis from undergraduates to graduates within the University, it will be interesting to see how much effort is made to save it.[29] The degree in Archaeology and Anthropology may never have been approved under such a policy, but at the moment it appears to be under no threat. The decline in undergraduate applications noted by Mills (2003a) has not yet occurred at Oxford, the intake remaining consistent at around twenty-five a year, but it is possible that the current enthusiasm for archaeology is maintaining the buoyancy and keeping anthropology afloat.

The last fifteen years of anthropology's century has seen two major administrative changes. The creation of the School of Anthropology and Museum Ethnography, composed of IBA, ISCA, and the Pitt Rivers Museum, has already been mentioned. Then, in 2000, as a result of a university-wide re-organisation into academic divisions, the School of Anthropology was placed in the Life and Environmental Sciences Division, and the Faculty Board of Anthropology and Geography was abolished. The School now consists of the ISCA which incorporates COMPAS, the Pitt Rivers Museum, and the Institute of Human Sciences. Whereas membership of the Life and Environmental Sciences Division suited anthropology, because it had a considerable voice within it, other subjects soon found the arrangement unsatisfactory. Further re-organisation in the first year of its second century placed the School of Anthropology in the Social Sciences

Division, but the subject starts that century with a far greater presence than might have been imaginable during certain periods of its first.[30]

Reflections on Oxford's global links

Finally, a few words should be included in this Introduction about the Appendix, although Wendy James, who was responsible for organising the session from which she has distilled this memoir, has provided its own introduction. Those who attended on the morning of Sunday, 18 September were treated to a fascinating array of comments, memories, and anecdotes from a range of past Institute members drawn from all over the world. It seemed an enormous pity to restrict such fascinating material to those there, so it was decided to include a report on the session as an appendix to this history volume – a place where it obviously belongs. Regrettably the demands of space have meant that only highly abbreviated versions of the various talks can be included and much of the richness of the originals has been inevitably lost. Despite that it adds an invaluable, personal touch to some of the chapters that have gone before.

Acknowledgements

I would like to thank Simon Bailey, the Oxford University Archivist, both on my own behalf and on behalf of most of the contributors to this volume. His assistance has been invaluable in putting this history together. I received helpful comments on a draft of this Introduction from many of the contributors. Gina Burrows and Rohan Jackson were invaluable at every stage of the organisation of the workshop, and Heather Montgomery and Mike Morris, the Tylor librarian, uncomplainingly checked references for me. As always, any sins of commission or omission remain mine.

Notes

1. Of the twenty-four contributors to Volume VIII of *The History of the University of Oxford: the Twentieth Century* (ed. B. Harrison, Oxford, 1994), just six did not hold Oxford University posts. This is exactly the same proportion as originally in this volume. It might be noted that since the original typescript was submitted in early 2006, David Mills has moved from Birmingham University to Oxford and a Preface by Alan Macfarlane has been added.
2. After a gap of two-thirds of a century, the return to a holistic anthropology now seems on the cards. For arguments in favour of this, see *Holistic Anthropology*, edited by Parkin and Ulijaszek, a volume which arose from another session that

formed part of the centenary conference and is to be published at the same time as this one.

3.   The occasion was marked by the publication of *The General's Gift* (see Cranstone and Seidenberg 1984). A great deal has been written and published about different aspects of the Pitt Rivers but a full length history remains overdue.

4.   The only evidence I have seen for its existence is to be found in a short notice in *Anthropological Review*, 5 (1867): 372–3 (see Anon 1867). The range of its interests was very wide and it heard papers on such diverse topics as 'The influence of Wyclifism on the national development,' 'The principles of war' and 'The statistics of crime'.

5.   In a sense the Pitt Rivers Museum was to Oxford anthropology as, in the decade of the 1890s, the Torres Straits expedition was to Cambridge anthropology.

6.   Few, if any, candidates sat the Supplementary Subject. The syllabus was large and equivalent to taking a full second FHS.

7.   Tylor reported yearly in the *Gazette* the numbers who attended his lectures. In 1900 it was three to ten members of the University and three to eight non-members. In 1901 the figures were five and ten (the latter being described as 'lady students and visitors'), and the following year there were fourteen altogether (*OUG* 1900–01: 672; 1901–02: 650; 1902–03: 548).

8.   These dates are the years in which undergraduates were first admitted to read for the degrees. The actual establishment of the degrees was in each case a year earlier; 1969 for Human Sciences and 1991 for Archaeology and Anthropology.

9.   There is a problem here because there is some doubt about what constitutes a 'department'. The title 'Department of Ethnology' is found in the *Oxford University Gazette* in 1891 where an authorisation for the Chest to make a payment to it is published (*OUG* 1890–91: 503). The name is used regularly from then on but it was not until 1935 that its statutory status as a department was confirmed. This was in the context of the appointment of Beatrice Blackwood to a Demonstratorship in Ethnology. She had previously held this post in the Department of Anatomy but in order for her to take up the post in the Department of Ethnology it had to be officially recognised and assigned to a Head of Department, the Curator of the Pitt Rivers Museum (I am grateful to Frances Larson of the Pitt Rivers Museum for this information).

10. The name was changed from 'Department' to 'Institute' in 1940 at the request of Radcliffe-Brown and by permission of the Chest and the General Board (see Mills, this book). The problem mentioned in the previous endnote concerning the status of the Department of Ethnology is even worse with reference to Social Anthropology. The permission to use the name, for it to have staff, premises and a budget did not give it official existence as a department, as Marett found out in 1935 when the Registrar and the Secretary of the Chest firmly stated it had not (see Davis, this book). James notes the same difficulty many years later (see this book). The problem became particularly acute in the late1970s, when it was found that the Professor, while responsible for the Tylor Library, had no such duties with regard to the Institute. Between 1978–90 the Institute was managed by a committee of the Anthropology and Geography faculty board. In the latter year general supervision of the Institute (or, in the words of the *Statutes* 'make provision for the lighting, warming, water supply, and cleansing') reverted to the Professor, but only five years later the *Statutes* were changed again so that 'a director' would be responsible for these duties. For much of its life, the Department/Institute was what is called a 'unit of academic administration' and nobody seems to know what its present status is.

11. The house in Keble Road had been home to Warden Spooner of New College who had been one of the most vehement critics of the proposed FHS in Anthropology in 1895.

12. One such new department was that at the Oxford Polytechnic, to become Oxford Brookes University in 1992. Anthropology had started there in the 1960s at what was then a college of technology, teaching anthropology for London University external BA's. Following the college's re-foundation as a polytechnic in 1970, modular courses in both social and biological anthropology, representing up to half a degree, were introduced. The department has grown greatly in the past decade, but from the start it was a sort of colony of Oxford University anthropology, with a large proportion of the staff having obtained their doctorates at the latter. I am grateful to Renate Barber, who was there through the earliest days, for this information.

13. Perhaps it was a case of sour grapes, but I heard E-P more than once declare that anthropology was not an undergraduate subject.

14. This was Paper II, entitled 'Ecology, Economics, and Technology', although the various Diplomas put their own gloss on it in the regulations. Thus, in both Prehistoric Archaeology and Ethnology 'Ecology and Technology' was added in parentheses, for Human Biology 'Human Ecology', and for Social Anthropology 'Ecology and Economics'.

15. Although, as in the case of Social Anthropology, it appeared to be a 'Department' it was in fact a 'unit of academic administration'.

16. For many years new members were elected at the AGM and this item on the agenda often resulted in long and heated debate on the qualifications of certain candidates. After a while this method proved so unworkable that the Committee of the ASA took over the vetting of candidates and putting before the AGM, for ratification, its list of approved new members. This usually speeded up the process, but occasionally the AGM would dispute a name. In the 1980s I can remember someone who is now a well-known social anthropologist having her candidature challenged on the grounds that her doctoral supervisor was an archaeologist.

17. There was a small but steady flow of people who were accepted to read for the Diploma who had no formal academic qualifications but whose careers were judged to provide a suitable background. For example, someone without a first degree whom I taught in the 1960s had been a qualified UN interpreter and has since gone on to be a distinguished anthropologist.

18. In practice members of the Institute took the chair in turn. My personal view is that, during those years, the Institute operated well, both administratively and socially.

19. One person with whom I broached the subject muttered something about a 'poisoned chalice'.

20. These posts were offered respectively to Harvey Whitehouse and Inge Daniels.

21. The archaeological curators were also included as part of ISCA until the further re-organisation in 2000 when they became part of the School of Archaeology.

22. Thomas Penniman, who had succeeded Balfour as Curator of the Pitt Rivers Museum in 1939, retired in 1963. The post was then held by Bernard Fagg (1963–75), who was followed Brian Cranstone (1976–85).

23. An obvious example is Aberdeen which got rid of all its anthropologists. In the long run, this may have been to anthropology's advantage as the subject was resurrected there in 1999 with a department of its own.

24. This period also coincides with the existence of the Management Committee of Social Anthropology and it may be difficult to separate out the effects of one from the influence of the other. I was involved and may be too close to have a proper perspective on what was going on. With that caveat my personal view is that during this period the Institute worked well but there was a lack of academic leadership. This does not mean that no one who occupied the rotating chairmanship was capable of providing such leadership, just that we did not think

it our job. We were all brought up in a tradition that subscribed to the idea of individual scholars getting on with their own work, and the idea of research agenda, research groups and research programmes (except on a very small scale) was alien to us. In retrospect, and given what was beginning to happen in the wider academic world, this was myopic but luckily did no lasting damage and the situation was rectified before the processes affecting academic funding started to bite.

25. The Diploma is also a consolation prize for those who fail the final of the M.Phil. The papers sat for the M.Phil. Qualifying Examination, taken at the end of the first year, are the same as those for the M.Sc. and Diploma.

26. Very few people now register for or complete an M.Litt. which has almost become a 'failed' D.Phil. This is a pity because, as James (this book) relates, some very substantial pieces of work, later published, were submitted for it.

27. This was said to result from the prejudice of the then Secretary of State for Education who among other things objected to the use of 'social science' in the title on the ground that it was not a science. In fact, the SSRC was lucky to survive in any form given the government's prejudice against 'social science'.

28. Although the input to this degree was originally intended to be equally weighted between the two subjects, in practice the emphasis has tended to be on archaeology. This has resulted as much from student preference as the respective commitment by archaeologists and anthropologists. Two years running I gave talks to groups of pupils from comprehensive schools brought to Oxford under the Sutton Trust scheme to encourage applications from such schools. They were nearly all more interested in archaeology than anthropology and their interest in the former had been aroused by the television series *Time Team*. Robert Parkin tells me that this fits with his experience of interviewing candidates for Archaeology and Anthropology for Keble College – most applicants have an archaeological interest, many having been on 'digs'.

29. Among other things colleges are showing an increasing unwillingness to admit undergraduates to read for Human Sciences. The size of the intake, but not the demand, has dropped from forty-four to twenty-six over two years. I suspect this is a response by colleges without tutorial fellows who can teach the subject to the externally imposed reduction in college fees paid for undergraduates.

30. Rodney Needham died in December 2006 while this book was in press.

Chapter 1

ORIGINS AND SURVIVALS:

TYLOR, BALFOUR AND THE PITT RIVERS MUSEUM AND THEIR ROLE WITHIN ANTHROPOLOGY IN OXFORD 1883–1905

Christopher Gosden, Frances Larson and Alison Petch

Any form of history is produced from contradictory motives. On the one hand, the historian attempts to be true to the past; to understand people and processes in their own terms, trying not to see events as the inevitable precursor of what we now know was to come. We should obviously resist the temptation to see the past purely through the lens of present concerns. On the other hand for history to be something more than a dilettante exercise it *should* speak to the concerns of the present, providing perspective and contrast to what is happening today. In this chapter we shall sketch the intellectual, institutional and personal situation of a nascent anthropology from 1883, outlining a very different organisation of knowledge to that which exists now and a university system that reflected this difference.

Anthropology at Oxford grew up initially as a minor element within the natural sciences, themselves still a small fraction of the University compared to classics and humanities. There were fierce debates on the benefits or otherwise of specialisation and the sciences generally grew up within the University Museum, partly designed to promote a unity of scientific endeavour. This is the past we shall attempt to understand in its own terms. But we shall also make two points of present relevance, both of them speculative. A theme running through the earliest history of anthropology within Oxford is the repeated failure to set up an undergraduate degree with anthropology as a major component. When these attempts did succeed, first with the establishment of the Human Sciences degree in 1969 and then with Archaeology and Anthropology in 1991, anthropology was set within a broad disciplinary compass which included variously the biological and social sciences, classics and archaeology. We wonder whether the birth of anthropology here, within a series of complex links to other disciplines, influenced its long-term history, making this eventual outcome less than an accident? Secondly, we are intrigued by the echoes resonating between a pre-disciplinary anthropology in the late nineteenth century and a situation developing in the present where disciplinary boundaries, so sharply erected and policed in the middle of the twentieth century, are starting to loosen up. We shall return to this point briefly at the end of this chapter.

The period from 1883 when Edward Burnett Tylor took up his post as Keeper of the University Museum until 1905 with the formal start of the Diploma in Anthropology is a complex and ambiguous one. George Stocking referred to Tylor and his contemporaries as part of an 'epistolary anthropology', engaged as they were in letter-writing campaigns to bring ethnographic information from the colonial periphery to the metropolitan centres of academia. We shall refer to this period as one of material anthropology, letters being just one of the material supports for the early discipline. Anthropology in Oxford was built around the Pitt Rivers Museum, founded in 1884 as a department of the University Museum, which was completed in 1860 and designed to provide a single home for the natural sciences (physics, chemistry, biology, zoology and geology, to use present day labels).

Research and teaching were carried out within the Museum, and for both these activities the collections were crucial, expanding quickly beyond the 20,000 objects originally given in 1884 by the eponymous Pitt-Rivers. These collections consisted of anthropological and archaeological artefacts, supplemented, in 1886, by the transfer of material in the University and Ashmolean Museums deemed to be ethnological. The collections were painstakingly developed by Henry Balfour (with some new material also coming via Tylor). Balfour was employed initially for a year in 1885 and was made the first Curator of the Pitt Rivers Museum in 1890. The expansion of the collections at the Pitt Rivers Museum, through letter writing, travel and conversation, provided then the foundation for anthropology in Oxford and provides now a well-documented means of accessing this history. Balfour especially used the collections to give the Pitt Rivers Museum a separate status from the University Museum and this in turn helped differentiate anthropology as a subject within the broader ambit of the natural sciences. Balfour continued to use the Museum as both a

Figure 1. Edward Burnett Tylor circa 1880. Copyright Pitt Rivers Museum, 1998.267.88

research laboratory and as a political tool within the University until his death in 1939.

The museum's collections also remind us that anthropology was a much more permeable and participatory discipline in the later nineteenth century than it is now. Material flooded in from all over the world sent by hundreds of people and often accompanied by letters or other forms of documentation pertaining to the use and history of the objects concerned.[1] On numerous occasions artefacts were sent which had been solicited by either Tylor or Balfour, so that accompanying written material was penned in response to specific queries. The complex network of people (most of whom were not professional anthropologists), objects and written documentation helped form anthropology at the time and had a lasting historical impact. For those interested in these histories now, the museum and its collections represents a rich source of evidence which can indicate the social and material networks existing from the late nineteenth century onwards. Before taking a glimpse at some of this evidence, let us first outline some details relevant to Tylor's arrival in Oxford, the position of anthropology within the natural sciences and within the University Museum.

Earliest Anthropology and its Context at Oxford

Oxford, during the nineteenth century, saw a shift from a single core curriculum, taken by all students and based around classics, to various specialised degrees in which eventually no trace of Greek, Latin or ancient history remained. From 1850 it was possible to take new subjects as part of the BA, each with their own board of examiners, and these included natural science, law and modern history (Curthoys 1997: 352). Taking natural sciences generally added an extra year to the degree, which explains its relative lack of popularity. Also, the natural science degree lacked college tutors and fellows, and was mainly taught by readers and professors employed by the University and able to charge for attendance at their lectures. Generally speaking, attendance at science lectures was low, which made it more difficult to lobby for a greater provision for the sciences within the University. Through the 1850s Henry Acland (who became Regius Professor of Medicine) and others, including John Ruskin, lobbied for a central building for science facilities then spread throughout the larger and richer colleges. This campaign, which crystallised some of the key tensions between humanities and sciences at Oxford, was surprisingly successful, resulting in the building of the University Museum on land south of the University Parks. The Museum opened fully in 1862 and provided a setting where the major sciences of natural philosophy (physics), chemistry and physiology could each be provided with laboratory and teaching space, but also maintain some overall unity in accord with Acland's vision that the sciences should give an undergraduate a broad and comprehensive grounding in the working of the natural world (Fox 1997). The Museum was based around a central courtyard with each of the major subjects on the north, south, and west sides. The east side was left available for expansion. The teaching of natural sciences was henceforth based in the University Museum which had space for practicals within its laboratories and a lecture theatre.

Also, from the 1850s onwards Augustus Henry Lane Fox Pitt-Rivers started making a collection, first of firearms, in keeping with his early employment in the army, but soon branching out into ethnographic and archaeological artefacts. Influenced by moves to promote evolutionary thought after the publication of *The Origin of Species* in 1859, Pitt-Rivers ordered his collection so that it would exemplify general evolutionary principles, key amongst which was the movement from simplicity to complexity, which Pitt-Rivers felt to be fundamental to human history, as well as biological evolution. His collection was assembled not just for research purposes but also for education, especially the education of the working classes, and was exhibited first in Bethnal Green and then in the South Kensington Museum. Pitt-Rivers felt that each of these venues was unsatisfactory for his purposes and canvassed a range of options in an effort to find a more suitable location. He had a number of close contacts in Oxford, including Henry Acland and, especially, George Rolleston, the first holder of the new Linacre professorship of Anatomy and Physiology, based in the University Museum. Rolleston and Pitt-Rivers were personal friends who had travelled and done fieldwork together in Scandinavia and Britain (Chapman 1982: 192) and it was Rolleston who probably provided the initial impetus for Pitt-Rivers to consider Oxford as home for his collection. Rolleston died in 1881 before negotiations with the University could be completed. However, negotiations were taken up by John Obadiah Westwood, the Hope Professor of Zoology, and on 30 May 1882 the University accepted the offer of Pitt-Rivers' collection: 'That the offer of Major-General Pitt Rivers, F.R.S., to present his Anthropological Collection to the University be accepted ... It will be seen that the Collection, besides having great intrinsic value, which from the scarcity of the objects themselves must necessarily increase as time goes on, it is of very wide interest, and cannot but prove most useful in an educational point of view to students of Anthropology, Archaeology, and indeed every branch of history.' (Chapman 1984: 23).

Running in parallel with attempts by the science lobby in Oxford to convince Pitt-Rivers to give his collection to the University, was a campaign to attract Tylor to Oxford. Tylor had been awarded an honorary DCL in 1875, along with John Lubbock and others who promoted a Darwinian view of social progress. Rolleston's professorship was split up in 1884 and distributed between a Linacre Professorship of Human and Comparative Anatomy, the Waynflete Professorship of Physiology and a new Readership in Anthropology (Fox 1997: 688). This represented an expansion of anatomical and physiological teaching and research within the University, but also shows that anthropology was conceived of as contributing more broadly to comparative studies of the human body and its relationships to the environment. It is also fitting that Rolleston, a great promoter of ethnology during his life, opened the way for the first post in the subject through his early death. Tylor was given first the Keepership of the University Museum in 1883 and then the Readership in Anthropology in 1884, starting his first formal lectures within the University in that year. Thus the ultimate beginnings of anthropology in Oxford date to 1884.

Tylor's duties were ill-defined, but probably included some responsibility for all the collections and activities of the University Museum, of which the Pitt Rivers Museum became a part. A new building was constructed to house this collection on the eastern side of the University Museum court. Work on the new building only began in 1885, and so the initial unpacking of the Pitt Rivers collection took place in the University Museum itself. Tylor had no special responsibility for the Pitt Rivers collection, which was managed by Henry Nottidge Moseley, the Linacre Professor of Human and Comparative Anatomy (a post created out of Rolleston's old chair). Moseley was responsible for unpacking and arranging the Pitt Rivers collection, while Tylor had an oversight role as Keeper of the Museum as a whole. When the collection was transferred from South Kensington Museum in 1884, Walter Baldwin Spencer was employed to help with the process. He was a natural sciences graduate, closely associated with Moseley, and was later to use the experience he gained from this close contact with a wide range of artefacts to good effect in his groundbreaking fieldwork in central Australia. Spencer commented of this time:

> [I]t was the old Pitt Rivers collection that first gave me my real interest in Anthropology. ... I did a great deal of the packing up and it was intensely interesting to have Moseley and Tylor coming in and hear them talking about things. I remember well that Moseley seemed to know a great deal more than Tylor in regard to detail and, of course, after his experience on the 'Challenger' he could speak of many things with first hand knowledge but Tylor with his curious way which you may remember of every now and then as it were 'drawing in his breath' – I don't know how otherwise to express it – simply fascinated me. It was intensely interesting to a young man like myself and also a great privilege to come into such personal contact with two such workers. Of the two it struck me at that time that Moseley had the greater technical knowledge but Tylor the wider outlook (PRM MS Collections, Spencer papers, Box IV: letter 21, 24 September 1920).

In October 1885, Moseley wrote to one of his former tutees in natural science, and a contemporary of Spencer's, Henry Balfour, to ask whether he could assist for one year in the unpacking of the collection, for which he would be paid £100. This temporary proposal became the start of Balfour's fifty-three year association with the Museum ending with his death in 1939, aged seventy-five. Balfour had read natural sciences at Trinity between 1882 and 1885, much of the teaching for which took place in the University Museum, using its collections. He fitted well into the general natural science ethos within which anthropology was created in Oxford and was disposed to think of the Pitt Rivers collection as an extension of the insects, fossils, human skeletal remains, plants, animals and minerals that composed the collections of the wider Museum.

In this respect, his approach mirrored that of other anthropologists in the same generation. Alfred Cort Haddon, for example, was a trained zoologist whose ethnographic research in the field and at home was, in many ways, an extension of his work as a marine biologist (Herle 1998). Large, systematic collections of specimens, which could be assembled, classified and analysed in great detail, were at the heart of research in the natural sciences at this time. Haddon and Balfour saw themselves as observational scientists, bringing their skills to bear on evidence for human cultural diversity and history. Other anthropologists within the Cambridge School, such as William Halse Rivers and Charles Gabriel Seligman, worked in the tradition of experimental science, conducting sensory and psychological tests on their subjects in the field. Collecting and evaluating objects was fundamental to this form of anthropology, which was so preoccupied with gathering ethnographic 'facts'. Whether working in the museum, surrounded by artefacts and specimens, or amongst indigenous informants far from home, this 'intermediate' generation of ethnographers sought to create a science out of culture and its products.

Balfour's work was firmly focused on the Museum in Oxford from an early stage in his career. When Moseley was forced into early retirement due to ill health in 1887, Balfour took sole responsibility for the collections, lobbying Convocation for more salary for himself and greater financial support for the collection. Balfour was at pains to stress in various forms of extant correspondence that he, not Tylor, was in charge of the collection and that only he was genuinely knowledgeable about the specimens and their manner of arrangement. Balfour was almost unknown during the 1880s and still in his twenties, whilst Tylor was the best known anthropologist in the English-speaking world. It is unsurprising that people inside the University, and without, should have associated Tylor's name with the creation of the collection, but Balfour was understandably irked by the fact that his hard work was not properly recognised and acknowledged. Through the 1880s and 1890s Balfour pursued a vigorous and generally successful campaign in the University to lobby for more resources for the Pitt Rivers collection and to have himself acknowledged as its first Curator. He was recognised in this title in 1890 and from then on the Museum started to separate from the Anatomy Department and from the University Museum more generally. Balfour not only created a job for himself, but provided ethnology with an increasingly separate status inside the University. This status remained ambiguous and disputable through into the early decades of the twentieth century (see Rivière, this book), but the existence of anthropology as a subject was made tangible and real through the growing bulk of the Museum's collections and the activities carried out through these collections. The transfer of ethnological material from the University Museum and the Ashmolean in 1886 was another significant step in the University's recognition of anthropology.

In June 1895 anthropology's profile was enhanced when a statute was approved to establish a Professorship of Anthropology for Tylor, tenable for the duration of his readership. According to the *University Gazette*, Balfour gave his first official series of lectures to students during the Michaelmas Term of 1893 (although he had taught students informally prior to this date). He talked on the 'Arts of Mankind' and used objects from the Pitt Rivers Museum collections to illustrate his words. As Reader in Anthropology, Tylor had been giving a series of anthropology lectures every term since January 1884, but in 1893 his lectures were announced in the *Gazette* alongside Balfour's and another series devoted to physical anthropology to be given by Arthur Thomson, then Lecturer in (soon to become Professor of) Human Anatomy. A special notice announced that, while they were open to anyone who was interested, all these lectures were 'adapted to meet the requirements of Students taking up Anthropology as a Special Honour Subject' at the University (*OUG* 1893: 603). This teaching continued into the early twentieth century when Balfour and Thomson were joined by Marett, together becoming known as 'the triumvirate' by successive generations of anthropology students. These three remained responsible for the University's core teaching in anthropology until the 1930s.

Tylor led a petition to establish a FHS in anthropology at Oxford, which culminated in 1895, but Convocation rejected his proposal, something he felt bitter about throughout his life. John Linton Myres (1953: 7) remembered that Tylor

> resented the rejection of his project for a degree examination in anthropology. It was an unholy alliance he said, between Theology, Literae Humaniores, and Natural Sciences. Theology, teaching the True God, objected to false gods; Literae Humaniores knew only the cultures of Greece and Rome; Natural Sciences were afraid that the new learning would

empty their lecture rooms. And the arch-villain was Spooner of New College, whom he never forgave.

Anthropology could only be taken as a special subject within the FHS of Natural Science, and this remained the case for undergraduate teaching until the first intake of Human Sciences undergraduates in 1970. One can speculate as to why anthropology failed to gain final honour school status, but part of its failure must be due to Tylor's lack of ability to galvanise support within the University. Although a popular and respected figure in the institution, Tylor seems not to have viewed the University as the sole centre of his life – he and his wife Anna took the train to Anna's family home in Wellington, Somerset, at the end of every term, only to return at the beginning of the next term. They spent under half the year in Oxford. Furthermore, Tylor lacked a college base (common for readers and professors at this time) until he received an honorary fellowship at Balliol in 1903, which reduced his ability to link into significant parts of the collegiate University. This might have been significant in his failure to convince the body of college tutors that anthropology was a suitable subject for undergraduate study.

In 1896 Tylor was unwell and underwent operations on 16 and 30 June. He seems from this point to have suffered an increasing incidence of ill-health, and his papers mention visits to famous spas like Baden Baden. In 1905 diaries record that he needed nurses and was in an invalid chair. Stocking (1995: 62) describes this period thus:

> In 1896, Tylor suffered a serious illness, and although he published several important articles after that time, it seems to have marked the beginning of a general mental decline, which by 1904 had become quite severe. When Tylor was knighted in 1912, Clodd wondered in his diary if he was even aware of it, since 'he has been long mentally dead'.

To summarise a complex and ill-understood period for anthropology in Oxford up until 1905, the successes included the recruitment of Tylor in 1883 and the award of a readership and then a professorship in the subject; the astute use Balfour made of the Pitt Rivers collections in establishing a museum as a separate entity with himself at its head; the large interest in, and number of people attracted to, the collections was also a plus point and this is explored below. On the minus side, there was the failure to establish an undergraduate degree in anthropology. This disappointment took place at the end of the century when the previous consensus that anthropology was part of science was breaking down – Edwin Ray Lankester, Moseley's successor, was as hostile to anthropology as Moseley had been sympathetic and was probably more concerned with establishing his own subject of anatomy than finding common cause with others. From the mid-1890s onwards anthropologists began to explore common ground with classics and archaeology, which was to bring relatively rapid results in the setting up of the Diploma in 1905. Anthropology started shifting from the natural sciences (which Balfour continued to see as its natural place until his death) towards the humanities and then the social sciences. Insufficient attention has been given in existing histories to this shift or the importance of the University Museum, the home of science generally, in providing the initial base for anthropology.

Participatory Anthropology

The Pitt Rivers Museum formed the first home of academic anthropology in Oxford, but its influence spread far beyond the confines of academia. The Museum only took the form it did through the active participation of many amateurs who contributed objects and information to the collections. Between 2002 and 2006 the Relational Museum Project has been looking at the history of the Museum's collections and the people collecting objects for the period 1884 to 1945. We will concentrate on the people here, as our statistics provide considerable data on the networks of people and objects vital to the creation of early anthropology. Most of the people participating in anthropology were not themselves anthropologists, nor did they have pretensions in that direction. The basic data for the Relational Museum project come partly from the computerised databases at the Pitt Rivers, which contain information on the objects, their nature and provenance, who acquired them in the 'field' and their routes to the Museum. These databases have been the subject of statistical analyses to throw up patterns in this complex (and slightly difficult) set of data. We have also looked at various archives (mainly those held at the Pitt Rivers itself) which contain a mass of letters, notebooks, field diaries and other written observations, by Tylor and Balfour on the one hand, as they either collected objects themselves or tried to solicit them from others, or from their 'informants' who carried out the collecting, who were often colonial officers, missionaries, travellers or academics. We have concentrated on six collectors who had important and differing connections with the Museum – Pitt-Rivers, Tylor, Balfour, Charles (and Brenda) Seligman, John Hutton and Beatrice Blackwood – and we will report on these in detail elsewhere (Gosden, Larson and Petch 2007). The wealth of information within the Museum's databases on numerous collectors and donors was an unexpected bonus, and we shall focus upon these people here as means for understanding some of the broader patterns at work in anthropology at the time.

In the period up to 1945, 4,674 named individuals were associated with the Pitt Rivers Museum, either because they acquired material in the field that later found its way into the collections, or because they donated (or sold) material directly to the Museum, or simply because they had owned objects in the collections at some point in the past. Between them, they contributed over 200,000 objects, a very significant proportion of the Museum's 289,000 objects (an ever growing total, estimated in August 2005). There is also a certain amount of biographical information for about 40 percent of these people and good biographical details for over a third (most of these being the stereotypical white, middle class males who feature in *Who Was Who* etc.). Of the 4,674 individuals who were linked to the Museum, some 966 were definitely women, around 20 percent of the total, and most people – on the basis of the biographical details we have – were middle class or aristocratic. Collections came from a particular section of society, but this directly reflects those interested in, and able to participate in, anthropology at this time. The number of objects collected by each individual varies. Balfour himself, who travelled extensively in Europe, Africa, Asia and Australia, collected over 12,000 objects in the field (and donated more) making him one of the Museum's most prolific collectors. Other major collectors in this period include Beatrice Blackwood, Charles and Brenda Seligman, Louis Leakey, John Henry Hutton, Edward Horace Man and William Matthew Flinders Petrie. However, this list is somewhat misleading in that these are names well known in the history of anthropology and archaeology. Over 4,200 of the people we are considering were associated with fewer than 10 objects at the Museum, and many of them donated only one or two things. The major point to make is that people who were not academics accumulated the bulk of the Museum's collections, through a mass of activity and recording work. The teaching, publications, and research work done by people like Tylor and Balfour was greatly influenced by this mass of activity. Their personal legacy represents the tip of a well-submerged iceberg.

The people for whom we have biographical information had the range of careers that one might expect (although designation to a career is more difficult the further back in time one goes, when academic disciplines, in particular, were not formalised). Colonial servants figure prominently as do the various branches of the military (see Figure 2). One might have expected missionaries and clerics to have a greater representation, but maybe a small number of prominent missionaries, such as Lorimer Fison, have given us an inflated view of the role of missionaries in general. Archaeology is the single most represented profession, which may reflect the fact that archaeologists are inveterate collectors for whom objects are a natural adjunct to academic life. Overall, the structures of Empire were crucial to these people's movements and activities. Colonial servants and the military were deployed throughout the Empire, often staying in one place for extended periods and getting to know local cultures. Missionaries and travellers mapped their activities onto the overall structures of Empire and added to them in turn, with global networks of mission stations and outposts vital to activities of anthropologists.

Not only were these collectors and donors clustered in certain professions, they had more informal links through social clubs and academic societies[2]. The British Association for the Advancement of Science (BAAS), which is of negligible academic importance now, was central to many of the debates and controversies underway during the nineteenth and early twentieth centuries, including those on anthropological matters (see Figure 3). The natural science base of the discipline is once again displayed through a significant membership of the Geological, Linnean, Royal and Zoological societies, whose interests overlapped with those of anthropology in a manner we would now find surprising. Involvement in the Royal Geographical Society and the Society of Antiquaries was also common amongst this group. These society memberships create a set of networks that were dense, complex and overlapping and can be subject to greater analysis[3].

Finally, although there are broader sets of connections through colonial infrastructures and club membership, the collections of the Pitt Rivers are refracted through links to Oxford. Over a tenth of people collecting had an Oxford University degree (the Diploma students formed a significant element within this group after 1907). A greater number of people were resident in Oxford for some part of their lives or donated material through people who were living in Oxford. In the middle of this web of connections sat Tylor and Balfour, both clubbable men, who, in Balfour's case at least, travelled widely making the acquaintance of many throughout the Empire. Through conversations, the influence of lectures and publications and letter writing, these two solicited material for the Pitt Rivers Museum and the influence of Balfour especially on the nature of the collections is difficult to overstate. Their relationship with the collections did not just reflect their intellectual interests, but helped structure and direct them. Tylor and Balfour thought differently as a result of their collecting work and the investigations they made of the Museum's collections. Balfour, in particular, saw the Museum as something of a laboratory in which ideas could be developed, tested or discarded.

A number of broad conclusions are possible, the chief of which is this: histories of anthropology have been mainly internal to the discipline, looking at a lineage of influence from Prichard to Tylor, Frazer, Haddon and Seligman up to roughly the First World War, with

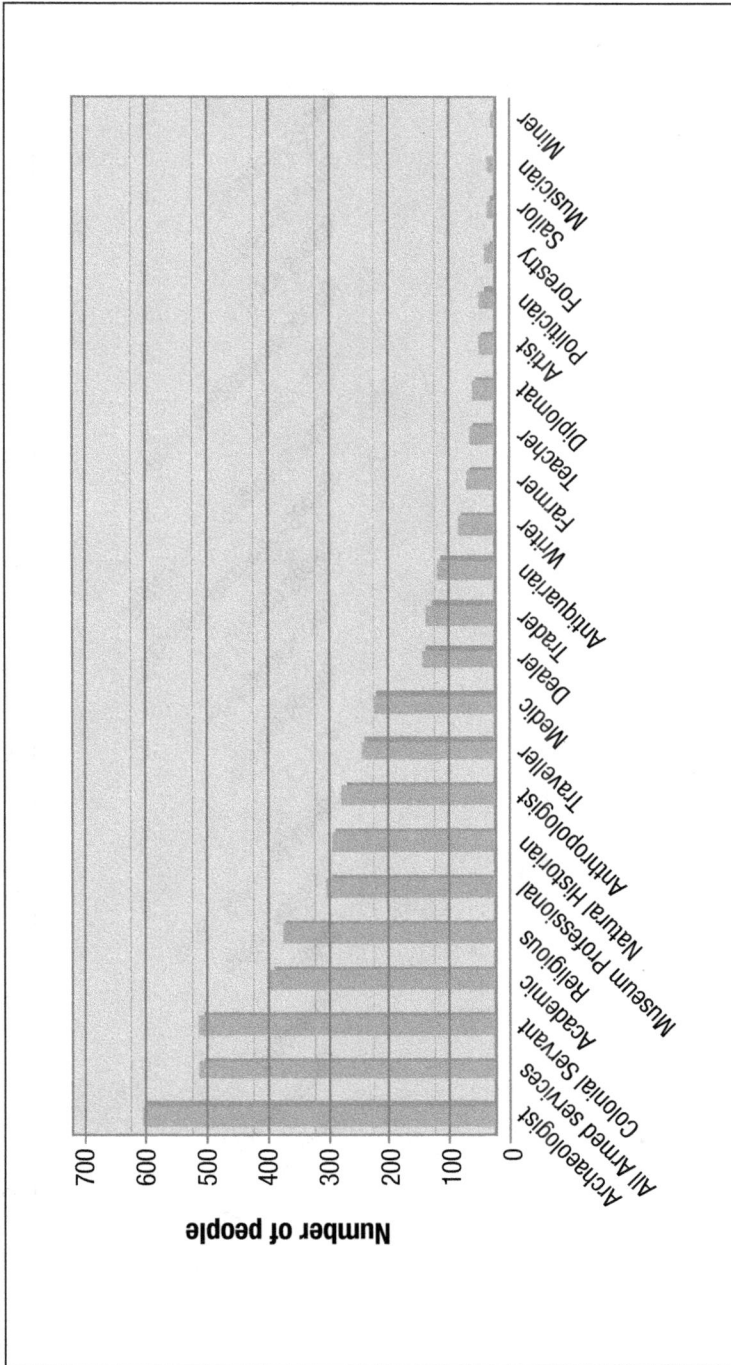

Figure 2. The careers of individuals associated with the collections at the Pitt Rivers Museum before 1945, where these are known.

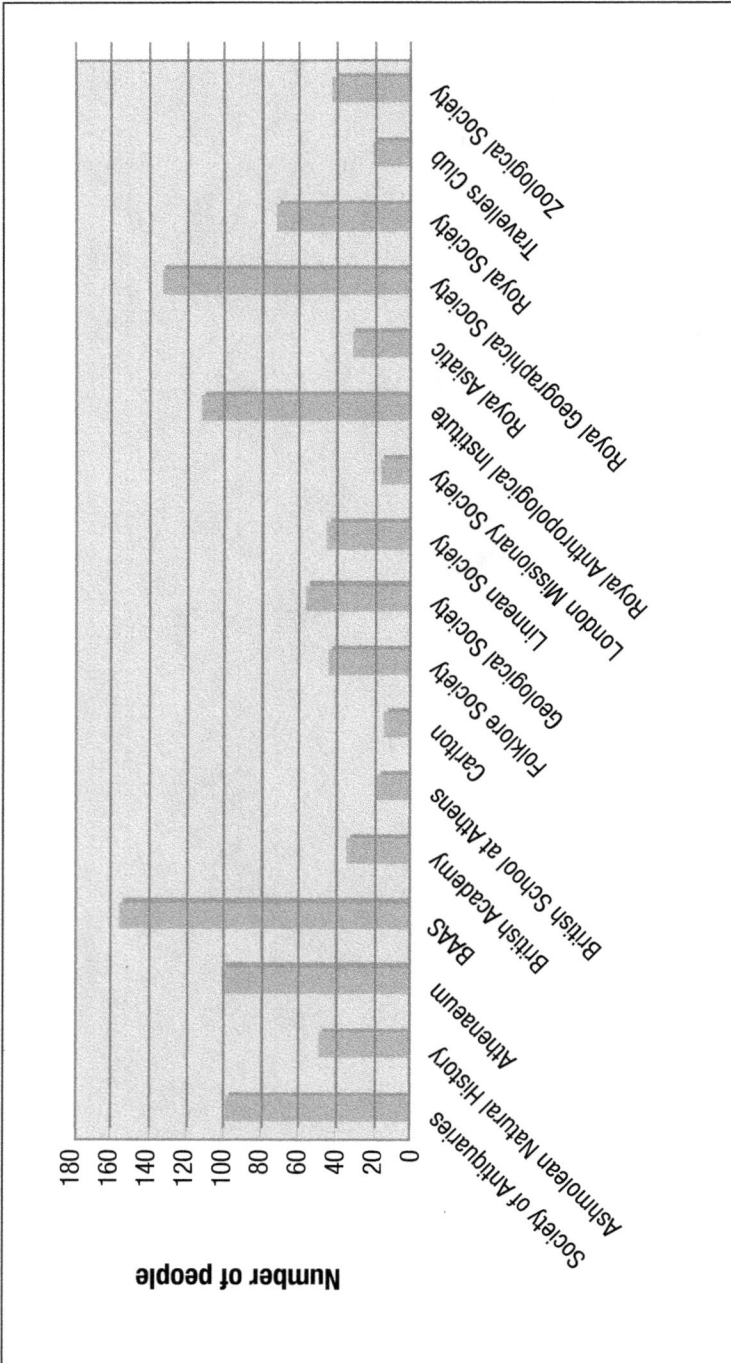

Figure 3. Club and Society membership for individuals associated with the collections at the Pitt Rivers Museum, where known.

Malinowksi and Radcliffe-Brown becoming ancestral to modern forms of social anthropology in the inter-war years (a disputed history, we know). But our data reveal a different and hidden history – in the period until at least the Second World War many people participated in anthropology, and anthropologists helped train members of the Indian Civil Service, administrators in the Pacific colonies, new missionaries and so on. This is not so much to make the old point that anthropology was handmaiden to colonialism, but to emphasise a broader historical conclusion that anthropology was a much more open and participatory subject, with the professionals a part of the whole, rather than the whole itself.

One vision for the future of anthropology is that it becomes participatory again, this time with a coalition between indigenous peoples and anthropologists, seeking common cause and sharing skills and perspectives. The older anthropology revealed here was antithetical in spirit to such late twentieth century views and had a fixed idea of who was able to participate, a vision which definitely excluded indigenous people. However, it might provide some social model for a new form of the subject, one that is more inclusive. Many at present within the Pitt Rivers Museum hope that this is the case. The second main conclusion returns us to material anthropology. The objects coming into the Museum did not provide people like Tylor and Balfour with the confirmation for conclusions that they held anyway; rather, they helped fashion and shape those conclusions. Without their collecting activity, their writing and teaching would have unfolded differently, because of the time they spent looking at objects and the myriad conversations and letters generated by those objects. Furthermore, objects did not merely flow through a set of social networks which existed anyway, but rather gave them shape and value, just as gifts do in many societies. Museum gifts came in many guises, either through the intriguing form of an object or the information contained in a letter or passed on in conversation. Such gifts needed reciprocal action, either as an acknowledgement in a publication or the knowledge that the object was on public display, or a more diffuse feeling on the part of the collector that they had contributed to an important intellectual and physical project, or had contact with well-known and influential people.

The Museum became the physical locus for teaching anthropology within Oxford and a means whereby Balfour in particular could lobby the University authorities for more people and resources. But the Museum was more than that – it represented a complex community of people and things which together helped shape academic knowledge and activity. As the powerful intellectual currents swirling out of mid-nineteenth century evolutionary thought died away in the twentieth century, the Museum came to be seen as an anachronism by a newly social anthropology. Changing academic fashions brought about the end of not just an intellectual tradition, but of an anthropology that was social in a different sense, in that it formed a society reaching out through national clubs and societies to the farthest reaches of the Empire.

Anthropological knowledge was the reciprocal creation of the new professionals and a wider network of collectors, a structure which more or less died with Balfour in 1939 and has been subsequently forgotten.

Discussion

The role and history of the Pitt Rivers Museum pose the question 'what was anthropology at the end of the nineteenth and beginning of the twentieth century?' Our answer has been that this past anthropology represents something of a foreign country when compared with later anthropology. Early anthropology in Oxford was firmly located within natural science, not only in teaching but also in the direction and spirit of research. The new professionals, of whom Balfour was the first real representative, were part of a broad community stretching way beyond the bounds of academia, most of whom might now be designated intelligent lay people. Balfour and Tylor's time was spent not just teaching and dealing with other academics, but participating in this community, which brought broad benefits but reciprocal demands.

The Pitt Rivers Museum was part of the larger entity of the University Museum, the home to natural science in Oxford. All lecturing prior to 1905 was carried out in the Museum and much thereafter, one suspects. Both Tylor and Balfour made constant use of the objects from the Museum's collections in their teaching. Here, Myres (1953: 6) remembers Tylor's lectures on anthropology:

> Soon after I came up in 1888, I was introduced ... to Dr Tylor, and attended his lectures in the Mathematical Room in the University Museum. The audience was small, mostly ladies ... Mrs Tylor sat in the front row, watchful for confusion among the specimens. 'Oh, Edward dear' she would say, 'last time, you said that one was neolithic.' But she did not prevent the conflagration when he demonstrated the fire drill, and his long beard became entangled with the bow.

Balfour in 1894, for example, ran a lecture series on 'Progress in the Arts of Mankind, particularly as illustrated by the Pitt Rivers Collection', and his teaching seems often to have been peripatetic, leading students around the Museum itself with a handful of miscellaneous notes and the collections themselves as his main reference (Wallis 1957: 786–7).

Figure 4. Museum House where Tylor lived after September 1883 and which housed the Institute of Social Anthropology between 1948 and 1952. It was situated in South Parks Road beside the Oxford University Museum of Natural History. The Pitt Rivers Museum's roof can be seen in the background. The figures are probably Tylor and his wife, Anna. Copyright Pitt Rivers Museum, 1999.19.6.

For Balfour, the decades of the 1880s and 1890s were generally positive and productive, as the collections grew and his own position became clearer. Tylor, by contrast, seems not to have enjoyed these particular decades as greatly. As usual for this period, the evidence is somewhat scant, but we can say with some confidence that Balfour viewed Tylor in an ambiguous manner – there is little doubt that the younger man had an admiration for Tylor's ability and achievements, but he did see that the route to his own preferment was to some extent blocked by Tylor's reputation, causing him to minimise any role Tylor might have had within the Pitt Rivers Museum. Tylor did collect and solicited material from around the world both for himself and for the Museum, not least bringing in the totem pole, the most striking object in the collections (Brown *et al.* 2000). Tylor lectured in the University Museum using objects from the Pitt Rivers collections on many occasions. He did have a role within the Museum and saw it as a crucial element within a developing ethnology. However, Tylor's time in Oxford may not have been that happy (although we have no statement to that effect from Tylor himself). We can surmise a slightly uncomfortable relationship with the energetic young Balfour. However, we can be much more certain of a difficult time with Edwin Ray Lankester (who was one of the more disputatious people within the University in the later nineteenth century). Lankester became Linacre Professor of Comparative Anatomy at Oxford in 1891. In February 1894 he wrote to the Vice-Chancellor to complain that Tylor (the Keeper of the Museum) had appropriated two rooms and the services of the attendant, Alfred Robinson. This dispute seems to have upset Tylor greatly, although it was eventually resolved.

Balfour had an intense, mutually constitutive relationship with the Museum: he created it, but was in turn created by it. The striking thing about Balfour is that, even during his first appointment to a very obviously temporary position whilst still in his early twenties, he started to engage in a single-minded and concerted campaign for the Museum, its collections and teaching activities. By contrast, Tylor's activities throughout his time in Oxford seem to lack any clear purpose or strategy beyond a general desire to see the role of anthropology grow. Maybe this is due to a generational difference in expectations as to what one can achieve within and through an institution. There are good grounds for seeing Balfour, and not Tylor, as the first professional anthropologist in Oxford. Tylor was fifty-one years old when he became Keeper of the University Museum and his most formative experiences and influential work came from the years he spent as a gentleman scholar, supported by family funds. His continuing influence within anthropology after 1883 derived only in part from his Oxford position and, although personally popular within Oxford, he was unable to turn this popularity into effective action on behalf of anthropology. Balfour was professionally employed from the age of twenty-two and lived his whole adult life within Oxford. His most striking successes came early, so that by 1890 he was recognised as Curator of the Pitt Rivers Museum, which was on its way to becoming seen as a separate institution as a result. He spent much of his energy and large amounts of his own money building up the collections and he kept up a stream of publications, most of which concerned the Museum's collections. By the end of his life, during the mid-1930s, Balfour was fighting an unsuccessful rearguard action against what he saw as the overly specialised anthropology of Radcliffe-Brown, which missed out the study of human anatomy, material culture or archaeology; subjects which to Balfour constituted the full breadth of anthropology. In this, Balfour represents the end of a broad natural science tradition harking back to Acland's vision of the 1850s. R-B, by contrast, was a prime mover in a decisive shift in anthropology's centre of gravity from the natural to the social sciences.

Compared to Balfour, Tylor was undoubtedly the more wide-ranging and original thinker. His major works, *Primitive Culture* and *Anthropology* represent a synthesis and condensation of social evolutionary thought, exploring both human unity (which had a biological basis) and social diversity, reflecting differentially effective cultural and historical processes operating across the globe. Savages were not in their primitive state as a result of mental or physical deficiencies, but because they lived in cultural arrangements that had not allowed the flowering of human capabilities. Marett (1936: 48) in discussing Tylor's view of cultural difference remarked that:

> Tylor's interest lay rather in the unity underlying this all too conspicuous difference. The need of his age was to proclaim that mankind is a many in one, with the emphasis on one. ... Natural Science, with no dogma to uphold, no axe to grind, was now prepared to state a case for human unity ... Tylor, in throwing in his lot as a student of Man with the new archaeology and the new biology, supported their demand for an indefinite allowance of time in which to find room for the human life-process to have run its leisurely course.

And as Tylor (1871, 1: 7) himself wrote:

> For the present purpose [of the book] it appears both possible and desirable to eliminate considerations of hereditary varieties or races of man, and to treat mankind as homogeneous in nature, though placed in different grades of civilization. The details of the enquiry will, I think, prove that stages of culture may be compared without taking into account how far tribes who use the same implement, follow the same custom, or believe the same myth, may differ in their bodily configuration and the colour of their skin and hair.

Tylor very much hoped that anthropology would take its place within the exact sciences and spent a considerable amount of time and effort in his later career on methodological issues. His 1889 article on laws of marriage and descent sought conclusions from data relating to 350 extant societies, using statistical analyses to create generalisations that were robust.

> Wherever anthropologists have been able to show definite evidence and inference, for instance, in the development series of arts in the Pitt-Rivers Museum, at Oxford, not only specialists but the educated world generally are ready to receive the results and assimilate them into public opinion. Strict method has, however, as yet only been introduced over part of the anthropological field. There is still to be overcome a certain not unkindly hesitancy on the part of men engaged in the precise operations of mathematics, physics, chemistry, biology, to admit that the problems of anthropology are amenable to scientific treatment. It is my aim to show that the development of institutions may be investigated on a basis of tabulation and classification (Tylor 1889: 245).

In terms of its methods and the reliability of its results, Tylor hoped that anthropology was a science, but he may well have entertained doubts that it was not, or else he would not have tried so hard to shore up anthropological means of proceeding.

In a more fundamental sense Tylor placed anthropology alongside biology in particular as an element of a broadly comparative science. Coming of intellectual age in the 1860s he felt the early force of Darwinism and believed in its ability to explain cultural arrangements. His desire to understand unity and differentiation sat comfortably within the Darwinian paradigm of biological science.

> How good a working analogy there really is between the diffusion of plants and animals and the diffusion of civilization, comes well into view when we notice how far the same causes have produced both at once. In district after district, the same causes which have introduced the cultivated plants and domesticated animals of civilization, have brought in with them a corresponding art and knowledge. The course of events which carried horses and wheat to America carried with them the use of the gun and the iron hatchet (Tylor 1871, 1: 8–9).

Tylor died before he had to cope with a sustained intellectual assault on his fundamental views from newer forms of anthropology and it was Balfour's ill fortune to appear as a reactionary at the end of his life, in a time of great change. However, intellectual currents move in mysterious ways, and the anthropology of the late twentieth century is seeing some return to the concerns of the late nineteenth. Tylor (and, to a lesser extent, Balfour) attempted to immerse cultural and social arrangements within the broad environmental background within which people lived. They did so, of course, through a belief in progressive change which we now find offensive, especially when allied to a condescending paternalism concerning the 'achievements' of other cultural forms. There is also a belief in science as a form of reason that is an antidote to unthinking traditionalism, which has been mainly swept away by a need to be critical of one's own approach and assumptions, and recognise them as deeply culturally embedded. Science cannot now be seen as a universal form of reason, but rather as a technically effective projection of western assumptions concerning cause and effect. These represent major differences to the programme of the late nineteenth century, which is unsurprising. Less expectedly there are some considerable similarities. Many are now calling for an understanding of people in their broader life-worlds, where processes of growth and change operate in rhythms that reverberate through the social and natural worlds. Indeed, many of the most interesting calls are to abolish the difference between the natural and the cultural (Ingold 2000, Descola 1994, Latour 1993, 2004). Material culture is again central to the thoughts of many, with both elements of the term – 'material' and 'culture' – exciting interest. Moves to do away with the conceptual frontier between the social sciences and the harder sciences would bring down disciplinary boundaries as they are currently constructed. While this is inherently unlikely for many reasons of finance and university politics – we are not all going to try to squeeze back into the University Museum – maybe new means need to be sought to bridge the divide that opened up during the first third of the twentieth century. As a discipline, anthropology, with its broad interests in the biological and the social, should be well placed to help lead new trends. Maybe the past holds part of the key to the future.

Acknowledgements

We would like to thank Peter Rivière for inviting us to contribute this paper and for commenting on an earlier draft. Thanks also to Jeremy Coote for his comments and to Norman Weller for his assistance. Any errors or infelicities are our own.

Notes

1.   Information on the history of the collections of the Pitt Rivers Museum and the early history of anthropology in Oxford more generally has been gathered through the Relational Museum Project funded by the ESRC (Award Reference:

R000239893) for which we are extremely grateful. The award was made to Chris Gosden and Michael O'Hanlon, and Frances Larson and Alison Petch have worked on the project as researchers. For more detail on the project see the website of the Pitt Rivers Museum: http://www.prm.ox.ac.uk/RelationalMuseum.html

2. Information on club and society membership is often lacking in biographical sources and could be extended through more detailed biographical work. As with much of the work carried out by the Relational Museum Project, we would see this as information to build on by future work rather than as a finished and finite piece of research.

3. Petch and Larson have been collaborating with David Zeitlyn, University of Kent, to create forms of network analysis which can start to reveal the multiple connections between people.

Chapter 2

THE FORMATIVE YEARS:

THE COMMITTEE FOR ANTHROPOLOGY 1905–38

Peter Rivière

The creation of the Committee

The formal recognition of anthropology at Oxford was achieved not through the efforts of anyone whom we might immediately recognise as an ancestor. It was the initiative of the Aegean archaeologist, John Myres.[1] He had a long term interest in anthropology, writ large, and played an important role in the subject during the first half of the twentieth century. He was directly responsible for drawing Robert Ranulph Marett into the anthropological fold when, in 1899, he invited Marett, at short notice, to give a paper to the Anthropological Section of the British Association for the Advancement of Science (BAAS) meetings in Dover.

Marett, a classicist, had already shown an interest in anthropology, for there was a close affinity between the two subjects at the time. In 1884 he had won a senior exhibition at Balliol College, got a First in Classical Moderations in 1886 and followed this up with a First in *Literae Humaniores* in 1888. In 1893, Marett, who by then was a fellow of Exeter College, entered and won the Green Moral Philosophy Prize, the subject for which, on that occasion, was 'The ethics of savage races'. Marett's BAAS paper, 'Preanimistic religion', constituted a critique of Tylor's theory of the origin of religion, which had been the orthodoxy for many years. He was later to write of this paper that it 'was in itself enough to secure me for all time a line, perhaps a whole footnote, in the text-book of every industrious compiler' (1941: 159). That this prophecy was wrong was evident even before he made it, but I shall return later to consider Marett's intellectual contribution to anthropology.

When, in 1902, Myres reopened negotiations on the matter of anthropology within the University, it was to Marett that he turned for help, and it was Marett rather than Tylor, who was ultimately responsible for the introduction of an anthropological academic qualification at Oxford.[2] The aim, however, had now become more modest, and instead of trying to establish anthropology as part of an existing Honour School the plan was to introduce a postgraduate qualification, a diploma, in the subject.[3] This was successful, although at considerable cost to the subject as a whole. David Van Keuren (1991) has argued, and I think rightly, that Marett's proposals were accepted because he was careful not to tread on other people's toes. The evolutionist anthropologists of the nineteenth century had seen the subject as covering all human development, social, cultural and physical, from palaeolithic savages to Victorian gentlefolk. Many aspects of such an all-embracing subject were, however, already claimed by others – historians, classicists, theologians and philosophers. If Marett could not trespass across their frontiers he had to colonise what was left, and that was 'primitives'. Van Keuren (1991: 54) has drawn attention to the 'contradiction between Marett's rhetoric and actions'; the former claimed anthropology 'to be the whole science, in the sense of the whole history, of man' (Marett 1912: 11–12) and the latter, for the sake of expedience dictated by the internal politics of Oxford University, severely limited anthropology's scope.[4] Thus it was that in 1905 Marett, when arguing for the introduction of the postgraduate diploma, stated that within the University at least, anthropology would limit itself to the study of 'past and present savagery'. This retreat from the study of human history cannot be seen as a purely Oxonian surrender, as much the same thing happened at Cambridge, where Frazer also proposed limiting anthropology to the 'crude beginnings, the rudimentary development of human society' (1913: 161). To some extent this narrowing of anthropology's domain chimed with the growth of fieldwork studies and the ethnographic tradition.

Negotiations to put in place a postgraduate qualification in anthropology were successfully concluded in the spring of 1905. On 23 May, a statute setting up 'a Committee for the organization of the advanced study of Anthropology, and to establish Diplomas in Anthropology to be granted after examination' was promulgated. The statute was accepted *nem. con.* by Congregation on 6 June and by Convocation on 14 June (*OUG* 1904–5: 536–7, 632, 661).

The Committee for Anthropology thus came into existence on 1 October 1905. On paper it had a heavyweight membership; the *ex-officio* members were the Vice-Chancellor, the two Proctors, the

Professor of Anthropology, the Linacre Professor of Comparative Anatomy, the Professor of Human Anatomy, the Waynflete Professor of Moral and Metaphysical Philosophy, the Wilde Reader in Mental Philosophy, the Corpus Christi Professor of Comparative Philology, the Keeper of the Ashmolean Museum, the Curator of the Pitt Rivers Museum. The elected members, who served for four years, consisted of two from each of the *Literae Humaniores* and Natural Science boards, and one each from the Modern History and Oriental Languages boards. There were also powers to co-opt further members both from within and without the University (*OUG* 1904–5: 728). In practice, few members attended the meetings, there being usually between six and ten members present, although at the 134th meeting on 11 March 1927, there were just two; Balfour in the Chair and Marett as secretary (Oxford University Archives (hereafter OUA) DC1/1/2). The Vice-Chancellor often chaired the meetings until 1928 after which he nominated a Pro-Vice-Chancellor to take his place.

The remaining clauses of the statute mainly concerned the duties and responsibilities of the new Committee for Anthropology, including arrangements for teaching, admissions and examination. Interestingly enough no University funds were to be made available for running the course and any monies had to be raised from fees paid by students and from any other sources (*OUG* 1904–5: 536–7).[5]

The early years of the Diploma in Anthropology

The Committee first met on 27 October 1905 and appointed John Myres, who was one of the Natural Science Board's representatives, as its first secretary (*OUG* 1905–6: 220). In good committee style the first thing it did was to appoint a sub-committee, the Sub-committee on Regulations. This was charged with the tasks of drafting regulations and a curriculum for the Diploma, and producing a syllabus of lectures. In very abbreviated form the resulting regulations required candidates to provide evidence of having been engaged in the study of anthropology for at least one academic year, and to satisfy the examiners in (I) Elements of Physical Anthropology; (II) Elements of Cultural Anthropology. The first was divided into three subsections which were (1) Zoology; (2) Palaeontology; and (3) Ethnology. The second consisted of four subsections: (1) Archaeology; (2) Ethnology; (3) Sociology; and (4) Technology. These were examined by four, three-hour papers and there were a further three hours of practical and oral examination (*Examination Statutes* 1906: 253–8).

The date for the first examination was set for 12 June 1906, but as no one entered for it, there is no evidence of what form it would have taken (*OUG* 1905–6: 754). When the examination was first sat in 1908, the papers were in Physical Anthropology; Ethnology; Archaeology and Technology; and Sociology (*OUG* 1907–8: 535).

In 1907, the year after the Diploma was introduced, the Certificate in Anthropology also made its appearance. It required a candidate to have been engaged in the study of Physical Anthropology or of Cultural Anthropology (Ethnology either with Archaeology and Technology, or with Sociology) for a period of three months.[6] The regulations also allowed for a candidate to earn a Diploma by an accumulation of all three Certificates over a number of years (*Examination Statutes* 1907: 259–60, 262).[7]

The statutes, regulations and syllabus, once in place in 1907, remained virtually unchanged for the next thirty-one years; until 1938, following the arrival of Radcliffe-Brown. One addition was the introduction of a 'Research Certificate' in 1930. This was basically to certify that someone was qualified to carry out anthropological research and it required either a Distinction in the Diploma or a statement from the candidate's tutor that 'his experience *has been* and his attainments *are* sufficient to enable him profitably to engage in a given piece of research' (*Examination Statutes* 1930: 339).[8]

The other change that occurred in anthropological teaching related to the subject as a Supplementary Subject in the Natural Science FHS. It had been included as an additional subject in 1885 and the attempt to make it a full subject in Natural Science in 1895 is dealt with by Gosden *et al.* (this book). During the period covered in this chapter the question of a FHS in Anthropology was raised twice more (in 1913 and 1930) without anything coming of it.[9] As a Supplementary subject it never appears to have attracted a candidate and its regulations had remained unaltered since its introduction. It was composed of seven sections: comparative anatomy; morphology of non-human primates; physical classification of races; prehistoric archaeology; comparative philology; development of culture; and practical work covering all the previous topics. In 1932 the course was changed so that a candidate covered the same syllabus as one of the three Certificates in Anthropology (*Examination Statutes* 1932: 107).

Throughout this period Physical Anthropology was housed in the Department of Human Anatomy, and Ethnology, Archaeology and Technology at the Pitt Rivers Museum, but the Department of Social Anthropology[10] had a more nomadic existence. It was initially based at Exeter College, but with a large rise in numbers in 1914 there was no longer room there and its base was moved to Barnett House, 26 The Broad, the Centre for Social Studies.[11] This move was made possible by a grant of £200 a year for three years from the Worshipful Company of Drapers of the City of London.[12] In 1920 Barnett House gave notice

Figure 5. Acland House, 40 Broad Street (fifth from right), was home to the Department of Social Anthropology between 1920 and 1936, in which latter year it was demolished to make way for the New Bodley. Anthropology shared the building with the Geographers. Copyright Bodleian Library, University of Oxford, MS.Top. Oxon.a.77, fol.12.

that it wanted the space back and the following year the social anthropologists moved into Acland House, 40 Broad Street, which it shared with the geographers (*OUG* 1914–15: 343–4; *OUG* 1921–22: 347; OUA DC1/1/2: 15 October 1920).

The structure of the course and the stability of its syllabus directly reflected the interests and longevity of the three people most heavily involved, Marett, Arthur Thomson, and Henry Balfour. Marett, who, as we have seen, was central in getting the Diploma in Anthropology recognised, was for many years, from 1907 to 1928, Secretary of the Committee for Anthropology (*OUG* 1907–8: 118; 1927–8: 679). The review on the conditions and requirements of anthropology that followed on Tylor's retirement in 1909 recommended that a Readership in Social Anthropology, at £300 a year, be created. Marett was duly elected to it on 1 July 1910 (*OUG* 1910–11: 414–15).[13] Thomson, an anatomist, was appointed to the new post of lecturer in human anatomy in 1885, at the time when the medical school was in its infancy. He became Professor of Human Anatomy in 1893 and the first Dr Lee's Professor, which went with a Studentship at Christ Church, in 1919. Tylor had early recruited him to give lectures in physical anthropology for the Natural Science FHS, which he did and for the diplomas until his death in 1935. Henry Balfour who was Curator of the Pitt Rivers Museum from 1891 until his death in 1939, covered the teaching of the technology and prehistory part of the course. He was a Research Fellow of Exeter in 1904–11 and from 1919 onwards, and received a personal chair in 1935. For Marett and Thomson the teaching of anthropology was what Marett described as an 'overtime employment' (1941: 272). Indeed from 1928, when Marett was appointed Rector of Exeter, much of his energy was necessarily directed towards the college.[14]

This triumvirate provided the permanent staff for anthropological teaching until the 1930s when death and retirement dispersed them. At the beginning, however, many other people were involved. For example, for Hilary Term 1906 lectures and instruction were listed under seven headings. 'General Anthropology' was provided by Tylor himself, while for 'Physical Anthropology', as well as Thomson, teaching was available from the Professor in Comparative Anatomy, the Professor of Physiology, and the Wilde Reader in Mental Philosophy, the last at the time being William McDougall who had been on the Torres Straits expedition. 'Geographical Distribution' was taught by the Reader in Geography (Halford Mackinder), and 'Prehistoric Archaeology' was in the hands of, among others, the Professor of Geology and Palaeontology (William Sollas), the Keeper of the Ashmolean (the archaeologist Arthur Evans), the Reader in Egyptology, and John Myres. 'Sociology' was covered by Marett, the

Figure 6. The triumvirate – Henry Balfour, Arthur Thomson and Ranulph Marett (seated from left to right). Standing behind, from left to right, are those who took the Diploma 1910, Wilson Wallis, Diamond Jenness, Marius Barbeau. This copy of the photograph was passed to Beatrice Blackwood by Wallis in 1939. Copyright Pitt Rivers Museum, 1998.271.11.

Corpus Professor of Jurisprudence, the Boden Professor of Sanskrit and others; 'Philology' by the Corpus Professor in Philology, the Jesus Professor in Celtic, and the Professor of Russian, who were joined a term later by the Professor in Chinese. Finally, 'Technology' was taught by the Lincoln and Merton Professor of Classical Archaeology and, of course, Balfour (*OUG* 1905–6: 302–3).

This impressive list of contributors was retained until the First World War, but after it there was a steady decline in those listed as being involved. By 1931, the teaching was almost entirely in the hands of the three stalwarts, Marett, Thomson and Balfour, with occasional help from Myres (*OUG* 1919–20: 19–20; *OUG* 1931–2: 398).

Diploma and research students

If we now turn from teachers to taught, a similar pre-Great War enthusiasm followed by a post-War decline may be noted. The first four students were enrolled for the course in Michaelmas Term 1907.[15] The following year this number dropped to two, but then rose steadily; three were admitted in 1909, seven in 1910, thirteen in 1911, sixteen in 1912,[16] and then a huge jump in 1913 to forty.[17] Not surprisingly the numbers fell back during the years of the First World War, the figures for the years 1914–18 being fourteen, seven, eight, nine and six. Following the war numbers initially rose with twenty-five in 1919 and eighteen the following year, but after that, and until 1937, the eve of R-B's arrival, the intake settled to around ten, with a high of fourteen in 1926 and a low of five in 1935.

When we turn to look at the background, or rather the 'qualification for admission', of the anthropology students, we find it to be very varied. Many of them are listed as having first degrees, but many are not, their qualification being stated as employment in some branch of the colonial service. Following the First World War, 'military service' was in many cases the qualification entered. One gets the impression that the entry requirements were not that rigorous, and perhaps that explains the relatively large proportion of those who enrolled failing to receive an award. The vast majority of the students were British, although there was a steady stream of North Americans, and in the 1920s and 30s a growing number of Commonwealth citizens, especially from India, appeared. The first student to register and one of the first two to obtain a Diploma was a woman, Barbara Freire-Marreco. There were nearly always one or two women on course each year. Exceptions to this are 1931, in which there was none, and two years of the First World War in which women outnumbered men; in 1915 there were four women out of seven students, and in 1916 six women out of eight.

The examination was sat for the first time in Trinity Term 1908. Of the four students, two of them got Diplomas that year, one got Certificates in Physical Anthropology and Ethnology and Sociology and obtained the Diploma by doing the other Certificate in 1909. The fourth one took the Diploma in 1910. In fact it was not unusual for students who entered on the course at the same time to sit the exam at different times. Nor did all students enter on the course in the same term, many, particularly those doing a Certificate, doing so in Hilary. A notable feature is how many of those admitted failed to receive any award, not it would appear from failing the exam but from not presenting themselves for examination. This reached an unprecedented level when of the forty people enrolled in 1913 only three are recorded as getting Diplomas and eight Certificates. Admittedly in that year many may have had more pressing matters on their mind.

If we now look beyond the general configuration of the anthropology students and take a brief look at some of the individuals who took the Diploma during its first thirty years, we get a better picture of what was going on. There were those who became fully or casually engaged in anthropology teaching. Early examples were Barbara Freire-Marreco and Francis Knowles, the first two people to obtain Diplomas (see frontispiece); the former, as a Research Fellow at Somerville, gave lectures on the Pueblo Indians of the American Southwest, and the latter became Thomson's Assistant in craniometry. Leonard Dudley Buxton, who obtained a Distinction in the Diploma in 1912, was appointed Demonstrator in Physical Anthropology the following year, Lecturer in Physical Anthropology in 1922 and Reader in 1927.[18] Mary Czaplicka, who took the Diploma in the same year as Buxton and went on to do fieldwork in Siberia, was paid by Brasenose College to lecture in the academic year 1918–19. In 1928, Beatrice Blackwood, who had obtained the Diploma in 1918, was appointed Demonstrator in Ethnology, but until 1936 she worked and taught in the Department of Human Anatomy, and it was only in that year that she transferred to the Pitt Rivers Museum. Even more eminent was Dorothy Garrod who took the Diploma in 1922 and went on to hold the Chair in Archaeology at Cambridge; the first woman to achieve such a position at either Oxford or Cambridge. Thomas Penniman, who had obtained a Distinction in the Diploma in 1928, became the Curator of the Pitt Rivers Museum when Balfour died in 1939. Others included E.O. James, who did the Diploma in 1916, was a lecturer at Cambridge and then professor of the history of religion and philosophy first at Leeds and then at London, and for the last years of his life he was chaplain at All Souls College. Then there were the Rhodes Scholars. E.A. Hooton, who was to become a professor at

Harvard and author of *Up from the apes* (1931), got a Distinction in 1912, and Clyde Kluckhohn, also to become a professor at Harvard, got a Distinction in 1934. This is just a sample and it would be possible, but of little value, to extend this list further. Rather it is useful to turn one's attention to another qualification that a number of students read for alongside the Diploma. For example, from the list above we might note that both James and Hooton obtained a B.Litt. contemporaneously with their Diplomas; the former in the same year and the latter a year after.

This leads to the question of the research degrees available at Oxford at the time. They were a junior one, a B.Litt. or B.Sc.,[19] or a senior one, the D.Litt. or D.Sc. There was no D.Phil., a degree that was not introduced into Britain and Oxford, until 1917.[20] The B.Litt./B.Sc. required a dissertation on a subject investigated during a course of special study or research and required eight terms residence although the actual course did not need to last longer than twelve months. For the senior degrees, the names of the candidates had to have been on the books of their college for twenty-six terms, and required the submission of original work in the form of papers or books. At the very beginning of the twentieth century, few people took a research degree. In the three academic years 1903–6, across the whole University, there were five B.Litt.'s awarded and eleven B.Sc.'s. Their popularity increased during the first decade of the century and in the academic year 1912–13, thirteen B.Litt.'s and seven B.Sc.'s were awarded. Between the wars, this number increased again and in 1930–1, there were fifty B.Litt.'s and thirty-one B.Sc.'s. In that same year eighteen D.Phil.'s were awarded.

To give some idea of the flavour of thesis topics, in the academic year 1908–9, three B.Sc.'s were obtained in Anthropology, all under the supervision of the Natural Science Board. They were Marius Barbeau on 'Some aspects of totemism of North-west America'; Francis Knowles on 'The correlation between the interorbital width and other measures and indices of the skull'; and Wilson Wallis on 'The conditions psychological and sociological of the development of the individual amongst peoples of rudimentary culture' (*OUG* 1908–9: 687–8). Although, as mentioned, the D.Phil. was introduced in 1917, it was not until 1936 that the first D.Phil. in Anthropology was awarded. It was to Max Gluckman for his thesis on 'The realm of the supernatural among the South-Eastern Bantu'. The second went to John Peristiany, two years later, for a thesis entitled 'The social institutions of the Kipsigi tribe' (*OUG* 1935–6: 874; *OUG* 1937–8: 969). Interestingly the former was submitted through the Faculty of Biological Sciences, the latter through the Faculty of *Literae Humaniores*.[21]

Another aspect that must not be ignored is field research. Students were involved in ethnographic fieldwork from the very beginning. Barbara Freire-Marreco, who, as mentioned earlier, was the first to enrol on the Diploma course, did fieldwork among the Tewa of Southwest United States. Francis Knowles, her contemporary was scheduled to join the 1910 joint Oxford – Cambridge expedition to Western Australia, although he withdrew for domestic reasons.[22] In 1908 Rivers recruited Arthur Hocart to the Percy Sladen Trust Expedition to the Pacific, and in 1911–12 Diamond Jenness was in Papua New Guinea.[23] Mary Czaplicka did field research in Siberia on which she based her work *Aboriginal Siberia* (1914),[24] and Beatrice Blackwood, in the inter-war years, went to Melanesia and North America. We should also take notice of the trio of Robert Rattray, Charles Meek and Francis Williams, who as colonial officers spent most of their time in the field. Rattray, who enrolled for the Diploma in 1909, but did not receive it until 1914, worked among the Ashanti. Meek, who was originally admitted in 1913 and then re-admitted in 1920, was a district officer in Nigeria. In 1947, he was appointed to a university lecturership, funded by the Colonial Office, to teach social anthropology to Colonial Cadets.[25] Williams, an Australian Rhodes Scholar, who turned to anthropology after military service, served in Papua New Guinea. All three produced very substantial ethnographic works. These are just a few examples, and some others will be given below. The important point is that field work was very much part of Oxford anthropology from its inception.

We can get some impression of what it was like to be an anthropology student before the First World War from an account by Wilson Wallis, an American, who read for the Diploma and a B.Sc. simultaneously, in 1908–10 (Wallis 1957).[26] He had a fairly positive view of the teaching he received for the Diploma although surprised by the degree to which students were allowed to go their own way. He appears to have taken full advantage of those whose lectures appeared on the anthropological list. He attended courses in human geography, comparative religion, psychology, Egyptology and lectures at the Medical School. His B.Sc. supervision was even more casual. 'One of my advisers for the latter [B.Sc.] was Henry Balfour; I never saw the other and I have since forgotten who he was' (1957: 785). He was obviously taken aback when Marett said to him '"If I could convince you that Balfour and I are dunces and everything we say is wrong, I think you might wake up and do something." This was not precisely in the academic tradition with which I had been acquainted' (1957: 786). At the same time he appreciated that 'Students were generally treated by dons as though on an equal footing with them, both intellectually and socially' (1957: 787).

This last comment is interesting and, at least in anthropology, the dividing line between teacher and taught seems to have been rather fuzzy. Hooton, who is referred to above, also gave instruction in physical anthropology in the year he was working for his Diploma and B.Litt. Perhaps the outstanding example of this is Arthur Maurice Hocart who read Greats at Exeter, where presumably Marett was his tutor, and studied psychology with Wilde's Reader in Mental Philosophy, William McDougall who, as already mentioned, had been a member of the Torres Straits expedition. On coming down, Hocart joined the Percy Sladen Trust Expedition to the Pacific, and when he enrolled on the Diploma course in 1914 he had completed six years of fieldwork. As a Diploma student, not only did he deliver a course of lectures on 'Problems in Anthropology' but also served as deputy Wilde's Reader for a term while McDougall was away. Hocart is a particularly striking example, but was by no means exceptional.

The colonial service course

As well as those reading for the Diploma and a B.Litt./B.Sc., there was also another large group of people for whom anthropological instruction was provided. They were the colonial administrators. The importance of anthropology in order to rule the various peoples across the British Empire had been noted by Lubbock in the opening words of *The Origin of Civilisation* (1870). By the end of the century this importance received increasing recognition. In 1896 the British Association tried to get the government to fund a Bureau of Ethnology for Greater Britain, using the argument that 'collecting information with regard to native races within and on the border of the Empire would be of immense use to science and to Government itself'. Although this proposal came to nothing, through lack of interest on the part of the Colonial Office and lack of any official funding, during the first quarter of the twentieth century, various further attempts were made to institutionalise the relationship between empire and anthropology. For example, in 1913 the RAI proposed the creation of an Imperial Bureau of Anthropology, and then in conjunction with the BAAS attempted to found a School of Applied Anthropology. A meeting, in February 1914, to discuss this was well attended; Balfour, Frazer, Haddon, Marett and Rivers were all there.[27] The meeting unanimously approved the resolution to develop the teaching of anthropology to colonial servants, and that this could be best achieved through the BAAS, the RAI and the universities with the co-operation of the Government, the Foreign Office, the India Office, the Colonial Office and Civil Service Commissioners. It was a bad date for getting anything done, but the project was revived again after the war (Temple 1913; 1914). At the meeting of the BAAS in Edinburgh in September 1921, a proposal to form an Imperial School of Applied Anthropology was accepted (Temple 1921). Nothing came of this either, and the responsibility for teaching anthropology to colonial officers was left with the four existing academic departments.[28]

It is not too surprising then that, in 1902, when Myres resurrected the case for the introduction of a course in anthropology, the value of an anthropological training for overseas government officers was one of the arguments put forward. This was put into effect when, in 1908, the Governor-General of Sudan, Sir Reginald Wingate, requested that Oxford and Cambridge provide anthropological training for probationers in the Sudan civil service. In its Report for 1910, the Committee for Anthropology confirmed that it was providing courses for Probationers in the service of the Anglo-Egyptian Sudan (*OUG* 1910–11: 415). The two universities then co-operated in an initiative to interest the India Office and the Colonial Office in similar training for their recruits. This paid off, and twenty out of the 1913 intake of forty-one were Public Service Officers, coming from India, the Malay States, and East and West Africa. From then on, and throughout the interwar years and after, teaching such probationer students remained one of the Committee for Anthropology's main tasks. Nor did the argument of anthropology's importance in the ruling of empire go away. At the end of Marett's reign in the mid-1930s, as Davis (this book) shows, the main reason why All Souls agreed to subvent the chair in social anthropology was the subject's value to colonial administration.

Marett and his anthropology

It is perhaps noticeable that nothing has so far been said about the intellectual development of anthropology in Oxford over these years. The reason for this is very simple: there was very little. Evidence for this is to be discerned in the almost total lack of developments in the Diploma course over a period of three decades. Some of the blame for this must be attributed to Marett. In the twenty-seven years in which he was Reader in Social Anthropology and virtually single-handedly in charge of the subject, nothing much changed in an academic sense. The anthropology that was being done at the beginning of the period was not that different from that at the end; most of the developments in the wider anthropological world – in particular the rise of functionalism and increasing specialisation within anthropology seem to have passed Oxford by.

Marett's reputation as an anthropologist has not survived and it would be difficult to identify any of his ideas that have had a lasting influence on the subject. Few people read his works today and then only for antiquarian or historical interest. But this was noticeable even by the mid-1930s. A majority of the contributors to the collection of essays presented to Marett on his seventieth birthday, *Custom is King* (Buxton 1936), makes no reference to his work and those that do, do so merely as a brief, preliminary courtesy. This is equally true today of the annual Marett Lecture at Exeter College which commemorates his name but rarely his ideas.[29]

The reasons probably lie in Marett's academic background and his introduction to the subject. When in the 1890s Marett became interested in anthropology little intellectual realignment was required, for at the time the distance between classics and anthropology was not great. Many classicists from Frazer to Gilbert Murray and Jane Harrison had turned in some degree to anthropology.[30]Marett, however, was unfortunate both in the sort of anthropology with which he became involved and in the timing of that involvement. The debates to which he contributed, and he was critical of both Tylor's and Frazer's ideas about the origin of religion, were those of the nineteenth-century evolutionists. Although there was considerable energy left in those debates, what was to become mainstream anthropology was already losing interest in them. By 1899 the Cambridge Torres Straits expedition had taken place and marked the beginning of the shift towards a fieldwork oriented future for the subject. Even the presence in Oxford of William McDougall, Wilde's Reader in Mental Philosophy, who had been a member of the Torres Straits Expedition, did not have any influence. By the 1920s this change had taken place, and there is a paradox here for while Marett chose to ignore this change in his own work, at the same time he was, from the beginning, fully supportive of and helpful to the students who undertook field research. For example, £258.3s was collected from the colleges to cover the cost of Jenness's field work in Melanesia, and there is little doubt that Marett was active in helping to raise this sum (OUA DC1/1/1: 16 June 1911).

This not to say that Marett was unproductive.[31] He wrote numerous journal articles for the non-anthropological readership and did much to bring the subject to the attention of the wider public. However, when we look at his serious anthropological volumes, those such as *The Threshold of Religion* (1909),[32] *Psychology and Folklore* (1920), *Faith, Hope and Charity in Primitive Religion* (1932), *Head, Heart, and Hands in Human Evolution* (1935) there is little in the way of theoretical development. This is also evinced by his textbook, *Anthropology*, which, first published in 1912, appeared in a virtually unchanged 11th impression in 1936.[33] Perhaps another revealing example is his article on 'Anthropology' for the 1929 edition of *Encyclopaedia Britannica*. Despite the fact that an encyclopaedia article should provide an overview of the particular topic, the only theoretical position referred to by Marett is the 'evolutionary method'.[34]

The titles of Marett's lectures in his last teaching year (1936–7), when he was Acting Professor of Social Anthropology because of Radcliffe-Brown's delayed arrival until 1937, also reveal a continued concern with origins. They included 'The religious basis of primitive culture'; 'Prehistory ancient and modern'; 'Primitive law'; 'Studies in the works of Tylor'; 'Primitive morals with reference to their religious background'; and 'Problems of cultural diffusion'.[35]

The end of the era and the questions of the future

Although the 1920s and 30s saw anthropology stagnate at Oxford,[36] increasing age and pending retirements brought about inevitable change. In May 1933 the Committee for Anthropology submitted to the Vice-Chancellor 'Memorandum on the position and prospects of Anthropological Studies in Oxford' which provides us with a snapshot of the situation at that time. The memorandum was submitted by exactly the same people who had helped create anthropology thirty years earlier; Arthur Thomson, Henry Balfour, Ranulph Marett, and John Myres. The memorandum pointed out that Oxford had been one of the first in the field as far as anthropology was concerned but now threatened to slip behind Cambridge, LSE and UCL. The subject had no official voice at any faculty board level, and was about to lose its premises which were earmarked for demolition to make way for the New Bodley. Finally, the Readership in Social Anthropology only existed because its present holder was a Fellow and Tutor (actually Rector by that time) of Exeter College, and his retirement was looming. The conclusion was that 'In the near future it seems likely that an independent chair or at least a whole-time readership will be necessary'. By the first week in June, Council had rejected the memorandum 'under the present financial conditions' (OUA DC1/2/2: 28 May and 9 June 1933).

The problem of premises and a replacement for Marett had, however, to be faced. The following year, in Trinity 1934, Council invited the Committee to prepare a paper on anthropology's most urgent needs for inclusion in its quinquennial bid to the University Grants Committee and which could also be used for an approach to the Rockefeller Foundation, the major player in the funding of social science. The result was 'Memorandum on work now done in

Anthropology in the University of Oxford with suggestions for its future development' in which a much more grandiose scheme was envisaged. The needs included a Professor of Anthropology – in any speciality – to act as head of the school and coordinator; readerships in Comparative Technology and Prehistoric Archaeology, and in Social Anthropology, five regional lecturers, a visiting readership, funds for graduate students, a publication fund and an ethnographic collection fund, equipment, and proper accommodation to include a library serving all specialities (OUA DC1/1/2: 6 June and 12 October 1934; DC1/2/3: Papers 217, 218).

The quickest result came in the form of five years' funding from the Rockefeller Foundation for a Research Lecturership with effect from Michaelmas Term 1935, to which to E-P was appointed.

In Trinity Term 1935, the General Board of the Faculties recommended the creation of a statutory Readership in Social Anthropology and asked Council and the Chest for their approval. This was duly given and the process of filling the post began. At this point All Souls College stepped in and its role in the founding of the chair and its other subventions of the subject is dealt with by John Davis (this book). It may be noted, however, that the filling of the readership was suspended early in 1936, then postponed, and finally, when the statute proposing the establishment of the chair was approved in June 1936, the readership was abolished (OUA DC1/1/2: 14 June 1935; *OUG* 1935–6: 382, 609, 714).

The other vital question that the 1933 Memorandum had raised was where would the Department be housed when it had to vacate Acland House? In the end room was found for it at 1 Jowett Walk in the same building as the geographers and where the geographers remained until they moved out this year (2005).[37] The building, known as Holywell House, had previously housed the Society of Home-Students which, in due course, was to become St Anne's. The move was made over the Christmas vacation 1936 and the new accommodation consisted of four rooms, one of which was a library. The Department remained there until 1948 when, because the geographers wanted more space, it moved to Museum House in South Parks Road, a very suitable location, having once been the home of Tylor.

The appointment of Radcliffe-Brown to the chair in 1936 and the changes consequent upon it are described by Davis and Mills (this book), but one aspect may be covered here, the demise of the Committee for Anthropology. At that Committee's meeting on 26 November 1937 it agreed to ask the General Board to establish a Board of Studies in Anthropology, mainly on the grounds that it could then deal with its own research student applications and examinations. The

General Board responded by proposing that there should be a Faculty of Anthropology and Geography and its Board would take over the functions of the Board of Studies for Geography, the Committee for Geography and the Committee for Anthropology. The Faculty was to be composed of two sub-faculties, that of Geography and that of Anthropology. This arrangement was approved and the Faculty came into existence at the beginning of the 1938–9 academic year.[38] The Committee for Anthropology held its last meeting on 10 June 1938. As an amazing symbol of continuity, John Myres, who had been the Committee's secretary at its first meeting in 1905, was in the chair (OUA DC1/1/2: 26 November 1937, 4 March 1938, 10 June 1938; *OUG* 1937–8: 969).

Notes

1. John Linton Myres (1869–1954), Wykeham Buxton Professor of Ancient History from 1910 and knighted in 1943.
2. Tylor's health by this time was not good and Marett was in many ways much better placed than Tylor to further the cause of anthropology within the University. The simple reason for this was that he 'belonged'. This may sound odd but Tylor never held an official college fellowship, the possession of which marked you as part of the establishment. Balliol did bestow an Honorary Fellowship on Tylor, but not until 1903. Not all academic posts became what is known as 'entitled', that is entitled to a college fellowship, until the 1960s and it was not until 1987 that entitlement had to be honoured and everyone holding an established academic post had to be offered a college fellowship.
3. There already existed a tradition of postgraduate diplomas. That in Public Health had been in existence since the early 1890s, and it was joined by a Diploma in Education in 1897, in Geography in 1900, Economics in 1903, and in 1905, the same year as Anthropology, one in Forestry.
4. Whereas Tylor had been concerned to introduce anthropology as part of the natural sciences, and had been defeated by the humanities, Marett was happy to accommodate himself to the latter, among whom, after all, he belonged. This does not mean that Marett saw social anthropology as a 'humanity'; he clearly saw it as a natural science (See James 2005: lxiv).
5. For example, Henry Wilde, who had endowed the Wilde Readership in Mental Philosophy, gave £100 in 1908 towards the work of the Committee for Anthropology (*OUG* 1907–8: 421).
6. This was changed in 1914 to two terms. The period for Probationary Colonial Officers was reduced at the same time to two terms for the Diploma and one term for the Certificate on condition that the Committee for Anthropology was satisfied that the candidate had paid sufficient attention to the study of anthropology (*Examination Statutes* 1914: 281).
7. In the 1960s, when I first arrived in Oxford, the Certificate was still available but was little more than a consolation prize for those who had not performed well enough in the Diploma examination.
8. Italics in the original.

9.   Further attempts to introduce a FHS in Anthropology were made by Radcliffe-Brown in 1937 and by Evans-Pritchard in 1948, as described by Mills and James respectively (this book).

10. The Vice-Chancellor had authorised the official use of this title in 1914 (OUA DC1/1/1: 1 May 1914). The previous year he had approved the use of 'School of Anthropology' in official correspondence (OUA DC1/1/1: 6 June 1913).

11. The present Barnett House, the Department of Social Policy and Social Work, is located at 32–6 Wellington Square.

12. This was followed up by £100 a year for a further three years, in 1918, 1919, and 1921. During late nineteenth and early twentieth century the Drapers' Company was very active in its support of technical and further education in Britain. At Oxford it gave money towards the building of the Electrical Laboratory and the Radcliffe Science Library, and to Hertford College, which fits with a letter from Marett to the Vice-Chancellor of 9 February 1924, in which he refers to the grant being obtained 'through the kind offices of the late Principal of Hertford [Henry Boyd]' (OUA UC/FF/228/1). However, there was also some direct association between anthropology and the Drapers' Company. In February 1914 an important anthropological meeting (see further below) was held in the Drapers' Hall. The following month the grant to Oxford anthropology was approved by the Company's Finance and General Purposes Committee (I am grateful to Miss Penelope Fussell, Archivist of the Drapers' Company, for useful information on this).

13. Except for Frazer's short-lived post at Liverpool this was the first position that had 'social anthropology' incorporated in its title.

14. Marett was due to retire from the Rectorship in 1936 but it was extended for five years and then in 1941, because of the Second World War, for a further year and renewable. He died still in office in 1943.

15. In fact it was slightly more complicated than that. The first student to be registered was Barbara Freire-Marreco and her date of admission is listed as Michaelmas 1907. An additional note states that she had started attending anthropology lectures in Easter Term 1905 and that the examiners deemed her 'to have entered as a Student of Anthropology as from M.T. 1906' (*Diploma in Anthropology*, I, 1907–50).

16. Marett (in Temple *et al.* 1913) provides a rather different set of figures.

17. It is not easy to explain this huge increase, although part of it can be accounted for by the rise in the numbers of probationers for the overseas civil services.

18. He was also Bursar of Exeter College. He died in 1939.

19. Postgraduate degrees with the title of Bachelor survived until the 1970s, when the misunderstandings this caused elsewhere led to their being retitled Master; the B.Sc. in 1971 and the B.Litt. in 1979. (The B.Phil., introduced in 1947, was also retitled that year except for that in Philosophy itself.) I seem to remember that the Junior Consultative Committee of the Subfaculty of Anthropology was quite influential in getting this change brought about.

20. For a study of the introduction of the Ph.D./D.Phil. into the United Kingdom, see Simpson 1983.

21. Both Gluckman, before he became Professor at Manchester, and Peristiany later held lecturerships at the Institute.

22. The Cambridge representative, being a certain A. R. Brown, proceeded alone (*OUG* 1910–11: 415; Stocking 1995: 310).

23. Numerous papers relating to Jenness's work in Melanesia and in Canada in 1913–14 are to be found in OUA DC1/3/1–2.

24. She was also one of the first to publish a popular and personal account of fieldwork, *My Siberian year* (1916), published by Mills and Boon.

25. Meek only stayed in post for a few years, and his successor was Mary Douglas (née Tew).

26. Unfortunately, the article consists of reminiscences recalled nearly fifty years after the event. Wallis is depicted in Figure 5, together with Diamond Jenness and Marius Barbeau. The last was a Rhodes Scholar from Canada, who went on to gain the title of Champion of the Order of Canada for his services to Canadian ethnography.

27. This was the meeting, referred to above, which was held at the Drapers' Hall.

28. The provision of courses for colonial officers was a valuable and important service for which there was competition between departments. For efforts by London University in collusion with the RAI to take it over, see Davis (this book).

29. An exception to this was provided by Harvey Whitehouse who delivered the 2005 Marett lecture entitled 'The evolution and history of religion', on the evening of the workshop, perhaps having heard my remarks.

30. This merging of anthropology and classics is all too obvious in the series of lectures Marett arranged in Michaelmas Term 1908 and published as *Anthropology and the classics* (1908). The contributors were Arthur Evans, Andrew Lang, Gilbert Murray, Frank Jevons, John Myres and William Warde Fowler.

31. For a list of his publications up to 1936, see Buxton 1936: 303–25.

32. This volume was republished in 1997 in a series, edited by B.S. Turner, entitled *The Early Sociology of Religion* (see Marett 1997).

33. This, of course, may have been the fault of the publisher. Not so long ago Oxford University Press insisted on bringing out, against the advice of more than one member of ISCA, an unrevised edition of a textbook which was years out of date, and not that good originally. The reason was simply commercial; it was still selling well.

34. For a very much more positive view of Marett's work and influence on anthropology, including the claim that he 'contributed to the reformation of British social anthropology', see Stocking 1995: 163–72. The point I am making is that how ever good Marett's arguments were in their own terms, those terms were those of an outdated anthropology.

35. There is an interesting contrast between Marett's titles and those of R-B in the following year which were 'Elements of comparative sociology', 'Sociological method', 'The comparative sociology of law and government', 'Comparative economics of simple societies', 'The comparative sociology of religion'. The most obvious point is the frequency of the word 'primitive' in one set and of 'comparative' and 'sociology' in the other. One needs, however, to be careful as Marett was not the only person at that time giving a course of lectures in which the word 'primitive' appears in the title; E-P was lecturing on the 'primitive mentality'.

36. Although Stocking in *After Tylor* does not use the word 'stagnation' in the text, it does appear in the index under 'Oxford, University of'.

37. The School of Geography is usually thought of as being in Mansfield Road, but the address for Anthropology is always given as Jowett Walk, the street that runs along the southern side of the School of Geography.

38. The Faculty of Anthropology and Geography continued until the end of the academic year 1999–2000 when, in a university wide re-organisation, it was abolished and the School of Anthropology became part of the Division of Life and Environmental Sciences. This, in turn, was abolished in 2006 and the School became part of the Division of Social Sciences.

Chapter 3

HOW ALL SOULS GOT ITS ANTHROPOLOGIST

John Davis

This story has three elements. The first is that in the 1930s social anthropology in Oxford needed strengthening: it was threatened by its internal weaknesses and by external competition. The second is that the University was unwilling or unable to help. And the third is that All Souls had a need to spend money on University purposes, and that men within the College thought social anthropology a worthy purpose. These three elements, combined, do not naturally react together to produce a Professor of Social Anthropology, and when you try to discover how events turned out as they did it is helpful to find a smoking gun: a letter, a note in a diary, that claims or shows a decisive role for a human being in shaping and manipulating the reaction. In the last months I have been searching for a smoking gun in the affair of the Professorship of Social Anthropology in Oxford: who was, who were the men responsible for ensuring that All Souls got its social anthropologist, and how did they do it? I have found only one apparently conclusive and non-circumstantial piece of evidence; it is a letter by R.G. Coupland to the Vice-Chancellor,[1] dated 29 October 1935. Coupland, who was Beit Professor of Colonial History and a Fellow of All Souls, wrote from his private address, marking the letter 'confidential'. 'My first moves *re* a Social Anthropology Chair have been successful', he said, and promised further developments (OUA, UR 6/ANT/2: 14).[2]

As you know, for its first fifty years or so in Oxford, anthropology was maintained by a combination of dedicated voluntary effort, a very small grant from the University, and a steady though hardly substantial income from fees paid by a limited number of students, by a few missionaries who thought the course was vocational, and by a rather larger number of Colonial officers on probationary or refresher courses. In the 1930s it was clear that the social anthropologists and their discipline were in a precarious position. Social Anthropology had not got a separate and established bureaucratic existence. Marett's readership was personal to him and although he had tried to insinuate a Department of Social Anthropology into the formal structures of the University, the Registrar[3] and the Secretary to the Delegates of the Chest[4] unmasked him in 1935.[5] The Committee for Anthropology in Oxford, which was the visible and bureaucratic entity, with a Minute Book and a Proctor to attend its meetings, was a broad and occasionally disputatious body.[6] That became completely apparent in the latter part of the decade, in discussions whether the Professor should be of *social* anthropology, or of anthropology unlimited.[7] The broader title would open the field to a wider ranger of academics. Lack of unity was also apparent in the attempts by Le Gros Clark to expel physical anthropology from the Anatomy department in the Museum (OUA, DC1/1/2: Minutes of the 168th meeting)[8] and of Major Mason, the Professor of Geography, to recover space from social anthropology.[9] Closer to heart, Balfour continued to resist the attempt to amalgamate the libraries of social anthropology and of the Pitt Rivers Museum.

On a wider stage, the London School of Economics Department of Anthropology, under Malinowski, was clearly outstripping Oxford as a centre for the development of ideas and methods in the discipline. The University of London was trying to recapture some of the teaching for colonial officers, given an expected increase in the number of trainees.[10] The Principal of the University[11] was in touch with the Colonial Office to offer comprehensive instruction from UCL, LSE, the School of Hygiene and Tropical Medicine as well as a number of lesser institutions, and in June 1932 he and Professor Coatman[12] sent syllabuses to the Colonial Office for a course to begin the following February. On 30 October *The Sunday Times* carried a puff for a course at the LSE: Lord Lugard would open it; it would be a useful qualification for those wishing to join the colonial administrations overseas. In the Colonial Office Tomlinson[13] and Bevir[14] minuted the file: it was wrong to imply that the course had Colonial Office approval; most participants would be officers on leave.[15]

If the Oxford colleagues were in any doubt that London anthropologists were attempting to poach their vocational colonial officer students, a further approach by the Royal Anthropological Institute under the presidencies of Lord Raglan and then J.L. Myres should have made the threat quite clear. The RAI had set up a committee on training for colonial service, and now offered to co-ordinate provision from all branches of anthropology in the country; they wrote to ask that an official from the Colonial Office should attend a meeting to identify the proper subjects; no one was sent. The Office asked whether Oxford and Cambridge were party to this approach; Bevir was unimpressed when told that the RAI Council was certainly representative, even if (it was implied) the committee of the RAI responsible for the approach was not.[16] The proposal died partly because the Colonial Office was satisfied with the service it got from Oxford and Cambridge, to the point of being protective against interlopers; and because the RAI was not of one mind. Malinowski for instance took pains to call on Tomlinson in the Colonial Office to explain his dissidence, and to propose a more appropriate course of real anthropology. A firm letter to Myres from Bevir implies that principals, professors and presidents should deal with the Secretary of State (effectively, with Bevir) rather than try to subvert particular officials in the Office (NA CO 323/1133/3, 15 July 1931).[17]

A third threat to Oxford social anthropologists came from a major effort in the University to establish a centre or institute for African studies. This was a proposal in 1929 [18] from an *ad hoc* committee chaired by H.A.L. Fisher, Warden of New College, with a mixed London and Oxford membership, including General Smuts, Leo Amery[19] and Philip Kerr.[20] Within Oxford the committee included the heads of Balliol, University, Hertford, All Souls and Exeter. Two other members were Fellows of All Souls: J.L. Brierly,[21] and Reginald Coupland whom you have already met with the smoking gun. The only overlap with the Committee for Anthropology was Rector Marett of Exeter.

The Fisher–Smuts–Amery scheme came to naught. In the Colonial Office, Bevir noted that the proposal was 'not popular' in Oxford, by which he presumably meant, unpopular with the administration (it was proposed that the new Africa institute should have independent endowment administered by the Rhodes Trust), if not with those who already taught the Africanists on the Probationer and refresher courses. Moreover 'the appointment of a Director who dabbled in politics would be an end to any value in the proposals so far as we were concerned' (NA CO 323 1069/6: Bevir, Minute 21 January 1930). By March, Kerr was writing to the new Secretary of State (Lord Passfield, *aka* Sidney Webb) to reassure him that they had been at pains to be 'non-Party and non-viewy'; and Passfield himself minuted that Kerr in conversation had denied rumours that either Smuts or Lionel Curtis[22] were to be the first Director. The scheme failed partly because of opposition in the Colonial Office, partly because the Rockefeller Foundation, which already supported the International Institute of African Languages and Culture in London, failed to turn up trumps.

Even though the scheme failed it was clear there was a recurrent background threat, an academic scramble for African studies, for which the social anthropologists had to watch rivals who were backed by a number of the Oxford notables, who were in turn backed by elements of the London imperialist intelligentsia.

The final threat was demographic. The remarkable men of the establishing generation were reaching the age of retirement; they had been supported by minuscule grants and by their own dedication and resources. They had no obvious successors waiting in line. So, in 1935 Balfour was seventy-two; Marett was sixty-nine; the physical anthropologist Buxton, a close friend of Marett, Fellow and Bursar of Exeter College, was indeed a stripling at only forty-five, but he died in 1939.

From the 1920s the Committee for Anthropology had tried to pre-empt these threats: they produced schemes of consolidation and expansion for social anthropology which would create posts and facilities, would secure an establishment. As Rivière (this book) has pointed out, the schemes became increasingly grandiose, and (after the slump of 1932–3) increasingly unrealistic. The Registrar conveyed Council's regrets: the University simply had not got the money; the only sources were the University Grants Committee and Rockefeller. The University would include social anthropology in its bids to each of those. In the end, as Marett's retirement came closer, the question shrank: it was no longer, who will give us a proper establishment, but – who will teach the students in Trinity term 1936? Marett wrote trenchantly, and so did Penniman, official secretary of the Committee for Anthropology.[23] In the end this practical issue brought some return: however lowly anthropology ranked in the University's expansionist schemes, students were not to be 'left in the lurch'. On 8 June 1935, the General Board of the Faculties sent a draft Statute establishing a Readership in Social Anthropology to Council which referred it to the Curators of the Chest. They replied on 17 June that they were 'inclined to accept'. It would cost £200 p.a. in addition to Marett's £300 p.a. stipend. '... it was always contemplated that additional money would have to be found when the present Rector of Exeter relinquished his office as Reader if the Readership was to be continued. It is clearly impossible to abolish it, and the Committee therefore recommends the Statute for approval' (OUA, UR 6/ANT/2: 3, 4).

This was agreed by Council in June 1935, and communicated to the Committee: the Provost of Oriel [24] on behalf of Council would propose a new statute to Congregation in the autumn, to establish a Readership in Social Anthropology.

In October that year, presumably in response to an inquiry, the Registrar wrote to Reginald Coupland enclosing a copy of the

University's application to the UGC on behalf of social anthropology. His letter said 'It does not give exactly what you want but you will see that a minimum of £1,000 was proposed'. It seems fairly clear that after Council's meagre offering in June 1935, members of the Committee mobilised their friends and people of influence to try to improve the terms. Coupland had been a member of Fisher's committee for an Africa centre, but he was not especially associated with social anthropology. He had been a Fellow and Tutor in Greats at Trinity College until Lionel Curtis persuaded him to take the Beit Lecturership in 1912, and he became Professor of Colonial History in 1920, in succession to Egerton.[25] It was a remarkable change of direction for Coupland, who became a well-regarded historian and a renowned lecturer in his new field. In addition to academic history he wrote a number of books that set out to capture the imagination of general readers, and wrote technical reports for the Colonial Office. He was a friend of Marett, who had perhaps kept him informed of the vicissitudes of social anthropology. Above all, because he held the Beit chair in All Souls he had been in close touch not only with Curtis, but also with Amery (Secretary of State for the Colonies until 1929). Adams, Warden of All Souls, was also known to be sympathetic.

With his letter from the Registrar, Coupland began manoeuvres in College, and on 29 October he wrote his confidential letter to the Vice-Chancellor: 'My first moves *re* a Social Anthropology Chair have been successful, and I am hopeful'. However, it would take time because of the schedule of College meetings; he could push the issue in College, 'but you will understand how easily a project may be imperilled if its authors give the impression of rushing it' (OUA, UR 6/ANT/2:14, Coupland to Vice-Chancellor, 29 October 1935).

It would be desirable to delay the Statute creating the readership, or to delay making an appointment, until the College had had a proper time to go unhurriedly about its decisions. And indeed in January 1936 the committee to elect a Reader in Social Anthropology asked Council for permission to delay an appointment until the outcome of proceedings in All Souls was known. That is the smoking gun: it seems certain that Coupland was the prime mover, within All Souls, of a proposal to convert the Readership into a Chair.

Coupland's 'first moves' were to propose a motion for the College meeting in November 1935, signed by him and six others, a good mix of academics and men of the world, of young and old. They were Charles Grant Robertson,[26] Donald Somervell,[27] J.L. Brierly (see note 21), Richard Pares,[28] Harry Hodson[29] and Arthur Salter.[30] Coupland's motion proposed that All Souls should contribute £600 p.a.; the University might then increase its contribution by £100 also to £600, and the readership could be upgraded to a professorship at

£1,200. Rockefeller had not been forthcoming; the University had no resources available, and only Colleges could provide the money. All Souls should do so because 'the contacts of 'civilised' with 'backward' peoples is one of the primary problems of our time'; the creation of a Chair would give the authority and prestige necessary to organise essential research. This was an argument designed to appeal to the practical imperialists in College (Green 2005). Moreover the older Schools in the University were interested in anthropologists' findings,[31] and anthropology was a 'necessary part' of the expansion of new social sciences at Oxford. This argument was directed at the group in College identified by Green as 'scholars'. In addition Oxford was falling behind Cambridge and London which had been expanding, incurring heavy expenses: that reflected badly on a University which claimed to be the birthplace of modern social anthropology in Britain. With a Chair, Oxford should claim an increasing share of the increasing numbers of Colonial Service Probationers, whose instruction properly included social anthropology (All Souls College (hereafter ASC), MS 145; also Minute Book, 1926–45).

Coupland's paper is undated, but on 20 November it stimulated the two Bursars to produce a counter-proposal: Geoffrey Faber[32] and Ernest Woodward[33] argued that social psychology had 'more to offer' than anthropology: it was useful to the study of the classics, to medicine, and biological sciences, whereas social anthropology dealt 'more directly with native races than with civilised society'. Again, it is uncertain how much more than nodding the Fellows' acquaintance with the discipline and its usefulness to others may have been. Philosophers, not mentioned in the Faber – Woodward flysheet, were divided: younger men thought psychology an ally in their assault on entrenched idealism; older philosophers thought it threatening (Morrell 1994: 143).[34] Faber asked a young philosophy Fellow, Isaiah Berlin, elected two years before, to write a report on the state of psychological studies in Britain. It is a young man's paper, brilliant and candid, not very helpful to Faber's cause. In any case it arrived in June 1936 when all the decisions about anthropology had been made. Berlin argued that a professorship in psychology would be inappropriate in Oxford: the University then had neither personnel nor *matériel* to support him. On the other hand, a research fellow might lay the ground for future expansion. So, a carefully selected scholar, working in the sub-field of psychology of the senses, of memory and cognition, might be useful in developing the science. But Berlin's path to that conclusion (itself consonant with the views of younger Oxford philosophers) was so un-silent, indeed so scathing about the pitfalls and shortcomings to be seen en route that (in my judgement) he effectively undermined any support the proposal might have got in College.[35] Faber, acknowledging Berlin's work, seems to recognise this.[36]

Coupland's proposal was discussed at the College Meeting in late November, and was referred (together with psychology) to the College's Joint Committee on Finance and Research.[37] Other proposals were circulated in College before the Committee reported: Foster, a barrister, joined with Henderson, the economist, in a paper of 4 February 1936 to suggest Criminology. On the same date A.H. Jones circulated a paper suggesting Byzantine History. In this interlude between meetings C.W. Oman produced a dismissive note to the College: 'Savage Customs and Experimental Psychology' were equally remote from his interests. As a historian he thought that Mediaeval Archaeology was producing information of major importance: 'two centuries of English History are at the present moment being re-written almost entirely by the aid of Archaeology' (ASC, Minute Book 1926–45 (2), 28 November 1935).

In a later note (March 1936) Oman, reiterating arguments heard in College since the 1870s, and his own consistent views, provided principled opposition to the creation of any new chairs. He argued that the College was becoming too big: from its primaeval forty it had already grown to fifty fellows, and it would lose its intimacy if it grew any more. Moreover, he said, the new chairs and readerships had brought people into College in their riper years: they were less careful of the College's traditions. 'Whenever pressure is put upon it from some influential quarter on behalf of some interest ... the College has endowed new senior posts'. Rather than new chairs the College should give substantial support to scholars in important specific projects. He had already circulated his opinion of the brilliance and importance of British archaeology at the present time; it would also be extremely valuable to publish decent editions of the mediaeval chronicles. The College had financed the young E.F. Jacob's edition of Archbishop Chichele's Register, and that was a thoroughly admirable expenditure. In short, rather than expanding the fellowship, the College should use its surpluses to become a grant-giving foundation 'on a very large scale' for specific purposes (ASC, Minute Book 1926–45 (2), November 1935).

Oman was the only College conservative to put his misgivings in writing. He had been a Fellow since 1884, only Cholmondeley (All Souls 1874–1937) his senior. He seems to have got no support. The tide of chairs and readerships was in full flow, and was the conventional orthodox modernising solution to the College's problems. It was helpful to the expanding and under-financed University, and it transferred surplus money from College in a way which, given the tax allowances, was cost free (see below).

We have reached late February 1936, with the Research and Finance committee deliberating, and eager fellows producing abundant proposals for spending money. Let us break off the narrative at this point to introduce two further elements of background.

The first is that at this time All Souls was relatively rich, embarrassed by its income, and the University was in deficit (Dunbabin 1994: 642). The University accounts were apparently no more intelligible then than they are now, but the admirably expansionist and enterprising University needed more money to increase its coverage of academic disciplines (notably the social and natural sciences and some humanities such as English and modern languages). The University got a grant from the government through the University Grants Committee: in 1930 the Treasury agreed to an increase to £97,500; it relied on Rockefeller and Nuffield and other benefactors; it relied on fees and on the proceeds of its own endowments. A fourth source was the Common University Fund (CUF), reformed by the Asquith Commission of 1922, to direct money from the Colleges to University purposes.[38] A joint informal committee of University, Treasury and College representatives agreed in 1929 that this account would be in deficit for the foreseeable future, to the tune of £4–5,000 p.a. The University was armed with the Act, and also with a 1909 Opinion of Sir John Simon[39] which said that in certain circumstances the University was empowered to appropriate Colleges' surpluses. The Colleges agreed in 1929 to make an *ex gratia* payment of £27,000 to offset the 1927 deficit, but they resisted a general reform of the taxation system, as well as any expansion of it. Most Colleges were historically poor, following the war and the slump. All Souls, which was not, preferred to spend money on University purposes, in the expectation that if it did so at a notably generous level and in a constructive way, it might avoid appropriation.

Colleges had strong material interests to resist reforms or expansion of University taxation. They defended themselves with pleas of autonomy which was no doubt the acceptable face of College self-interest, but it was more than that. It was also a legal fact, and it was the basis of that freedom of action that allowed Colleges for instance to set up laboratories and experimental centres from the seventeenth century (Christ Church, Wadham, Balliol): it was a source of innovation and reputation.

All Souls was not strapped for cash. Like the other ancient Colleges it had relied on agricultural land for its endowment income, but in the later 1800s the College began to let land at rack rents (rather than leasehold) and income increased, at any rate until the agricultural depression of the 1890s. The College then developed land for housing on its Middlesex estates. It was at first its own developer, building roads[40] and drains, and selling leases on the enhanced land to builders: the growth of suburban north London in effect raised income from about £5 per acre to about £50, and the return on investment in roads and sewers at 5–6 per cent compared favourably with the 2–2.5 per cent on government securities. Kensal Rise was substantially developed by 1920, and as railway stations were built at Hendon and Stanmore, further College land was developed in Cricklewood, Queensbury, Kingsbury and Edgware. From about 1925, the College ceased to play an active role in development, and simply let land on long leases, receiving ground-rents.[41] By the 1930s the College was accumulating unspent income ('surplus') at a rate of about £30,000 a year. Bursar Faber predicted in a paper of 1932 that this was likely to increase. It was in some senses embarrassing. All Souls had had forty fellowships in 1858; after the great reforming commission of 1858, the College suppressed ten of them to provide stipends for professors. The awkward surplus continued, however, and the College then undertook a modest academic expansion, especially in the 1890–1920s, to absorb some of it. Nevertheless, a series of inquiries and commissions ostensibly about the future of Oxford University and its Colleges, from 1920 until the Franks report of the 1960s, all managed to include a few paragraphs on how All Souls might contribute more to the collective academic effort. Then as now a number of Colleges were sure that if they were endowed as well as All Souls was, they would do much better.

The College tried for a while to increase the number and the subjects of Prize Fellowships, and to include natural scientists. These experiments failed, and the College reverted to its annual election of two fellows after examination. The next strategy was to increase the number of research fellows and, as Oman noted, an increasing number of fellows were men who had come into College in their maturity, at age forty or fifty, and had not learned the College at an impressionable age. As well as electing research fellows, the College used its money to support teaching and research elsewhere in Oxford. All Souls Readers in Roman Law, in Statistics, in English literature for instance were paid by the College but were housed and did tutorial teaching elsewhere. Finally, the College improved conditions for fellows, especially in the 1880s: stipends increased, benefits improved, in an effort to make a fellowship a reasonable employment in itself, rather than a sociable adjunct to income from law, politics, or indeed inherited wealth.

The fourth way of spending money was on University purposes. Between 1930 and 1940, and including the tax paid to the Common University Fund, the College spent about 20–25 per cent of its income in this way, making grants to Bodley, supporting readerships and new disciplines, and creating new professorships. You have noted how the

Commission of 1858 imposed the suppression of fellowships and the creation of chairs attached to the College. Even this did not always go smoothly: the University understandably would have preferred a general subvention, freely disposable by the University, rather than a series of earmarked sums. The faculties tended to resent All Souls' corner in Chairs, and they agreed that the College should have no more professorial lawyers or historians: the Chichele Chair of Economic History, created in 1931, was a provocation to the faculty, passed off as a contribution to the development of social studies.

I think all fellows agreed that All Souls could not become again, as it had once been, a club for not very academic but well-connected young men. That granted, the College certainly contained a number of influential fellows who thought it should use its money to support scholarship in traditional fields, such as Byzantine history. Simon Green has characterised these as the 'scholars', favouring individual researchers contributing in a scholarly way to the advancement of knowledge. On the other side were those who thought the College had an opportunity to become something new – a research institute with a remit to study practical issues. They were practical men who wanted to solve pressing contemporary problems. Their experience and their interests were primarily imperial and colonial, and Green characterises them as 'imperialists', men with practical experience of power and administration, with imperial connections, who favoured a school of administration or government, broadly understood. They thought, to varying degrees no doubt, that good administration required reliable knowledge of the administered (Green 2005). It may be that imperialists in the University supported Coupland's proposal to All Souls: but Coupland (who was clearly on good terms with Warden Fisher of New College and with the Registrar and Vice-Chancellor) would not have emphasised this support in College: he needed to persuade the scholars.

The College supported in whole or in part the creation of seven chairs between 1900 and 1940.[42] Why was social anthropology included among them? Green suggests that the chair in social anthropology was a compromise between the interests of the College's 'scholars' and those of the 'imperialists'. Although their interests were worldwide, their connections were chiefly with India and with South Africa.

So far as India was concerned, apart from three Viceroys of India[43] known to fellows alive in 1936, the Fellowship included a number with direct experience of India, or born there to families with a tradition of public service: Charles Grant Robertson,[44] Patrick Reilly,[45] Hodson,[46] F. W. Bain,[47] M.L. Gwyer,[48] E. D. Swinton[49] and Rushbrook-Williams.[50] Penderell Moon[51] had just gone out of Fellowship. Even Oman had an Indian connection, though not through public service: his father had been an indigo planter.

In short, All Souls in the mid-1930s had a number of fellows and former fellows with a connection or an interest in Indian affairs, and were perhaps susceptible to the argument that systematic knowledge of the governed was an important source of stability and fairness in the government of India. They may also have believed (a slightly different issue) that the appointment of an eminent and authoritative Professor of Social Anthropology would help create such knowledge.

The College also had established interests in South Africa. Four fellows were recruited to work with Milner in South Africa during and after the second African war, and others joined the Round Table movement at a later stage. Sir Alfred Milner (1854–1925) was appointed Governor of Cape Colony, and High Commissioner of South Africa, in 1897, and remained in post until 1905. His was, says his biographer in the *ODNB*, the most turbulent pro-consulship in the history of South Africa. He was taciturn and self-assured, attractive to women (he did not marry until 1921), and an inspiration to Christian Britons. He was never a Fellow of All Souls but in South Africa he attracted an extraordinary number of able and committed young men who were devoted to his ideas, and convinced that they elaborated them in their own lives' work. The dozen or so young men who were part of 'Milner's Kindergarten' held posts of importance in the administration of post-war South Africa. In later life they achieved distinction in different branches of government and academic study. They founded a journal, *The Round Table*, which Milner supported with funds from the Rhodes – Beit trusts, and a movement: men dedicated to the idea of an open, quasi-democratic and Christian enlightened empire, who joined in conferences, groups, associations also called The Round Table, owing allegiance to London, but organising their own deliberations in the colonies and dominions. All Souls provided four of Milner's original dozen, who had worked under his spell in the crucial and devastating years in South Africa. These were Curtis himself, Robinson (later Dawson),[52] Brand[53] and Malcolm.[54] They were supplemented by younger fellows: Coupland, persuaded by Curtis to take the Beit chair in 1920, wrote for *The Round Table*, and was an assistant editor for a while. Hodson was assistant editor in 1931, and editor from 1934. Of other fellows, Dermot Morrah became editor of the journal in 1944, and Leo Amery was an occasional contributor although he seems to have been too canny, politically, to assume office in the organisation.

All this tends to suggest that in 1935 All Souls had a sufficient number of fellows who had a quasi-religious belief or a belief sustained by religion in the inevitability and justice of a British Empire, and in its mission to civilise the world and *perhaps* to create a peaceful and just world order. They had experience in particular of India and of South Africa, but were internationalist in view. For many reasons they too were susceptible to ideas that All Souls should contribute £600 p.a. to the stipend of a Professor of Social Anthropology.

The College met in March 1936 and received a report from the Finance and Research committee that the College should contribute £600 p.a. to the Chair of Social Anthropology, with an amendment from the Estates Bursar that the College should meet the superannuation costs for the entire stipend. The College accepted, as it also agreed that other rival proposals for expenditure should be considered at a later date. Marett wrote a letter to Adams, thanking him for his support; the Registrar wrote a letter congratulating All Souls on its regular generosity to the University, and in due course the Board elected A.R. Radcliffe-Brown to the new Chair of Social Anthropology in the University of Oxford. His successors have been welcomed in the College ever since.

That is the story: a 'department' in distress; a University unable or unwilling to help; a College with an urgent need to spend money on University purposes.

The College's decision was more or less unconstrained by considerations of economy: it did not have to choose among alternatives on financial grounds. And it did not choose after a balanced or 'objective' weighing of the intellectual values of the alternatives. Rather, it was Coupland who was able to find a proposal that would gain assent from the conventionally progressive modernising scholars (creation of chairs was a fashionable remedy), and from the practical men who wanted All Souls to be useful to government and administration. The Chair in short was a compromise between different interests within College, and was sufficiently attractive to secure a majority in support, against those who might well have doubted the aims and understanding of Coupland and his allies: as often happens, interest and expediency swayed the College.

People who support any proposal do so from mixed motives: it is very rare to find univocal majorities in colleges, governments, synods, international organisations. Coupland was able to create a majority from scholars who thought (questionably) that a Chair in Social Anthropology would provide greater understanding, or at least 'insights' into their own disciplinary problems,[55] and from those practical men who thought (questionably) that the Chair would bring greater justice and fairness to the lives of the administered peoples of the colonies and dominions. Of course, when we look at other peoples' histories, we are usually aware of the ambivalences and compromises and idealism that produce humdrum mixed results from the desire to do good and to improve the world. Indeed, I owe to my supervisor Lucy Mair the belief that any social study which does not compare people's intentions with the outcomes of their actions is unworthy of the name: irony and compassion should suffuse our work. But while we can achieve this with some ease when discussing other histories, it is a little more difficult in our own history. When we look back, seventy years later, on the decisions of 1935–6 we tend to think of ourselves as we are as if we had been *intended* by the far-sighted and wise men who allocated the funds. But they were not immune from ordinary shortsightedness and from mixed motives, and I hope I have shown that the Chair did not have an immaculate conception. I am sure that several of the influential men who ensured that there should be a Chair were taken aback even from an early date, by the personnel and the development of the discipline they had fostered.[56]

Postscript I

All Souls' support proved crucial in later years when colleagues attempted to capture the chair by dropping the 'social', opening the succession to future, physical, palaeontological and museum anthropologists.

It was at first of course a Readership to replace Marett and was therefore designated to be in social anthropology. All the discussions in College about the possibility of converting it to a chair retained the epithet *social* and you may imagine that Coupland's arguments would have been less persuasive if they had been advanced to support a generic anthropology. The possibility of it becoming such arose first in Hebdomadal Council in 1934–5 where a paper urged that in designating new chairs, the title should be as broad as possible. The University was expanding into new fields which were certainly exciting but perhaps also unstable. Where rapid development was to be expected it might be sensible to give a broad title to a chair, to allow for future shifts in emphasis, consequent on advances in knowledge. The Vice-Chancellor sent a copy of the paper to All Souls on 6 May 1936, to consider whether this might apply when the next, second, appointment was made to the Chair of Social Anthropology. The Warden replied on 1 June, after a discussion at a College meeting. The College, he said, appreciated the general arguments, and would accept the change if ever it should be recommended by Council. 'But, as you know, it has made its contribution because of the special interest in Social Anthropology and it trusts that this fact will be given due consideration in all future appointments to the chair' (ASC, MS 435).

There the matter rested until 1945 (when Radcliffe-Brown's tenure was coming to an end). The Secretary of Faculties wrote to Warden Sumner on 22 January 1946, asking for the College's view of a proposal from the Faculty, to change the name from Social Anthropology to simple Anthropology. Sumner wrote the next day to the chairman of the Faculty Board: 'No reasons are given for the proposed change. It would help the College in considering the matter if you could give me at any rate informally the main reasons why the Board desires the change'. J.N.C. Baker of Jesus replied on the 24 June: the Board of the Faculty of Anthropology and Geography had set up a small *ad hoc* committee to consider whether the chair should lose its specific 'social'. Radcliffe-Brown wrote to Warden Sumner that the proposal came in the first instance from the Professor of Anatomy, Le Gros Clark, and was strongly supported by Professor (Major) Mason, of Geography. R-B had opposed setting up the committee, and 'pointed out' that All Souls might withdraw its funding if the change were accepted.[57] The *ad hoc* committee reported to the Faculty Board in January, and was in favour of the change. The chairman of the Faculty said that at this stage neither the Board nor the *ad hoc* committee was aware that All Souls had already considered the matter in 1936, and had then declined to agree to the proposed future change in the title. When that became known, two of the *ad hoc* committee members withdrew their support from the proposal, and all the anthropologist members of the board voted against, with R-B abstaining. The votes were tied, and the chairman of the Faculty used his casting vote against the report. 'The position thus became rather ridiculous and a very long argument followed. In the end it was arranged to send on the report but to ask for the comments of All Souls'. I suppose it possible that in 1946 two or three powerful men on a Faculty Board might be able to impose their will to insist that a rejected report should nonetheless be forwarded, but it is unlikely to be a common occurrence in the twenty-first century. The main reason for the recommendation was to strengthen the subject generally, Baker said, 'but there are other reasons which I would prefer not to put on paper'. Baker and Sumner met on 29 January, but no record survives of what the 'other reasons' might have been.[58]

In March 1946 Warden Sumner wrote to the Secretary of Faculties stating 'that the College are specially desirous of furthering the study of Social Anthropology, that they consider that the proposed changes might result in less attention being given to Social Anthropology, and that they are therefore averse to the proposed change of title'.[59]

Postscript II

In addition to the annual subvention for the Chair, the College made a further annual grant to the Department of £300, from 1951 to 1967. Warden Sumner had suggested to Evans-Pritchard that, since the All Souls Lecturership in Law was coming to an end (it was converted to a Readership and the whole costs borne by the University), the College might use the sum saved for social anthropology: would E-P write a note on what the money might be used for. He suggested three purposes: a research lecturership which would allow the Institute's lecturers to conduct field research; a research studentship, to allow good students to continue after the B.Litt., and a publications subsidy fund. On Sumner's advice E-P proceeded with only the first of these, which was agreed at a College Meeting in June 1951.[60] The College thereafter paid £300 to the Institute, until the University took over the whole funding of the departments and faculties in 1967.

The College possesses an airmail letter from Godfrey Lienhardt, dated 7 December 1952 at Akobo, Upper Nile Province, in which he makes an interim report to the College on his fieldwork in Anuak country:

> The Anuak custom is to gather themselves around an important man, enhance his importance by doing so, eat everything he possesses and then desert him. I cannot afford to be thus deserted, so I am put to considerable expense in coping with the demands of self-appointed retainers; some of the funds provided by All Souls are therefore, indirectly, keeping rather truculent Nilotics in native tobacco, a curious charity for the College! ... I shall spend Christmas at the court of the King of the Anuak. He was one of Professor Evans-Pritchard's irregulars in the Abyssinian campaign (ASC, MS 435).

Acknowledgements

I acknowledge with gratitude the help of Dr Simon Green, paramount historian of All Souls; Dr Norma Aubertin-Potter, Librarian in Charge, ASC; Dr Juliet Chadwick, Librarian of Exeter College; Professor Wendy James; Dr Douglas Johnson; Professor Roger Louis, and Dr Peter Rivière. It will be clear after a page or two, how much I owe to the *Oxford Dictionary of National Biography*. While the Chair in All Souls is important to social anthropology in Oxford, to the holders and to the College, you should not let this get out of perspective: the people involved were busy, active in politics and academic life. By giving brief sketches of the lives of those principally involved, I hope to have conveyed the incidentality of the issues and events described here.

Notes

1.   A.D. Lindsay, Master of Balliol 1929–49.
2.   Lindsay noted he had replied, but his letter does not survive. Coupland's extensive papers in the Rhodes House Library do not include documents on this matter.
3.   Douglas Veale (1891–1973). Bristol Grammar and Corpus Christi, Oxford. He served in France and Belgium 1914–17, and then was a civil servant until his appointment in Oxford (1930–58).

4. Sir Arthur McWatters (1880–1965). Clifton, and Trinity, Oxford. He served in the Indian Civil Service, 1903–32. He was Secretary to the Chest – the senior financial officer of the University – 1932–46. In India he had been loosely associated with the Round Table: see below.

5. In 1914 the Drapers' Company had made a most munificent benefaction of £200 p.a. for three years. That was later extended for a lesser amount for a further three years. Marett ensured (or could not prevent) that they made their cheques out to the non-existent Department, and he opened a Chest account for it. The then Vice-Chancellor Heberden authorised a decree accepting the gift and the Committee for Anthropology accepted that the Reader in Social Anthropology should administer the funds (OUA, UR 6/ANT/2/17: Social Anthropology, Readership in.: Secretary to the Chest to Registrar, February 1935). The account remained when the Drapers' grant eventually came to an end. Marett used it for odd bits and pieces of income, paid for example 'a librarian' £15 p.a. to look after the Social Anthropology library. In 1935, when the Registrar noticed the anomaly, the account held a balance of £200-odd. It was all accounted for quite properly, annually, to the Committee for Anthropology.

6. Membership in 1933 was: the Principal of Jesus (A.E.W. Hazel), chairman, the Senior Proctor (T.R.F. Higham, Trinity); Professor Goodrich; the Keeper of the Ashmolean (E.T. Leeds, Brasenose); the Rector of Exeter (Marett); Professor Griffith; Professor Mason (Geography); Professor J.L. Myres (Wykeham Chair of Ancient History, 1910–39), Dr Buxton; Dr Sandford; Mr. J.N.L. Myres (Student of Christ Church); Mr Penniman (Secretary). It was the Professor Myres who was President of the Royal Anthropological Institute 1928–31; his son, also an archaeologist, later became Bodley's Librarian.

7. See below, Postscript I.

8. See also the chapter by Gosden *et al.* in this book.

9. Marett to McWatters, 28 November 1936. 'Mason, when agreeing to share a double perambulator between his large baby and my small one, did it, apparently, under the false impression that the latter would never grow bigger'; 'I cannot consent to be put off with two small dark rooms ...'. Exeter College, Marett Papers, L.VI.II, Miscellaneous correspondence 1929–42.

10. Warren Fisher, of the Treasury, wrote a Report in 1930 on the organisation of the Colonial Office. Among other recommendations, the number of officers should be increased, and their training arranged such that no more than forty trainees were present in any one University at any one time. That implied an expansion of provision, beyond Oxford and Cambridge. Fisher (Winchester and Hertford), became Permanent Secretary to the Treasury and the first official Head of the Civil Service in 1919; he was an unpopular and controversial figure, and appears to have been a semi-retired consultant by the mid-1930s (National Archives (formerly Public Record Office, hereafter NA) CO 323 l079/10: Warren Fisher Report).

11. Sir Edwin Deller (1883–1936), a lawyer and academic administrator, had been Registrar 1921–9, and then Principal, 1929–36.

12. John Coatman (1889–1963). Indian Police Service 1910–21, becoming Director of Public Information, etc. He was Professor of Imperial Economic Relations, London, 1930–4, and was about to become chief News Editor of the BBC, 1934–7. He was Director of Research in Social Sciences, St Andrew's, 1949–54.

13. Sir George Tomlinson (1876–1963). Charterhouse and University College, Oxford; KCMG 1934. He had served in the education branch of the Transvaal administration under Milner, and then in Nigeria (1907–28) with secondments to Ghana (Gold Coast); he became a Lecturer in Hausa at SOAS in 1928–30, and then worked in the Colonial Office, 1930–9.

14. After war service, in which he was badly gassed, Anthony Bevir (1895–1977), Eton and Hertford College, entered the Colonial Office 1921–39, was then in the War Cabinet Office, and became Private Secretary to successive Prime Ministers, 1940–56, (KCVO, 1952). He was admirable and generous but found Churchill's conduct of business too irregular and stressful, and was entrusted with the Prime Minister's patronage business (Colville 2004: 678).

15. It was customary for the overseas administrations rather than the Colonial Office to grant extra leave to officers who attended a course, and to make an allowance to cover some of the costs. The London proposals died the death with the slump. Tomlinson to Deller, 26 January 1932: 'Since we first started our conversations the financial position of the Colonies however has become such that it seems unlikely that we shall be having any extensions to our training courses' (NA CO 323 1079/10).

16. To complete the record, the Colonial Office also received an offer in November 1934 of assistance from 'the Institute of Anthropology of Edinburgh' whose letterhead stated it was 'Governed by the Standing Committee on Anthropological teaching in Scotland (as 'jointly established by the Council of the Scottish Anthropological Society and the Senatus of the Free Church College, Edinburgh)'. It was signed 'R. Kerr, Hon. Sec.' The Office was nonplussed, and inquired of Professor Hodson of Cambridge whether he knew anything about the Edinburgh signatories. The Office seems to have taken no further steps (NA CO 850 39/4 'Anthropology').

17. Perhaps Tomlinson, with his appointment to SOAS, was thought to have gone native with the academics in London.

18. Inspired by a remark of General Smuts in November 1929, that 'the future of Africa depended primarily on the British peoples' a number of people had attended meetings at Rhodes House. Fisher now sent a circular letter, dated 5 December 1929, setting out the proposal (NA CO 323 1069/6).

19. Leo Amery (1873–1955). Fellow of All Souls, 1897–1912, 1939–55. Correspondent of *The Times* in South Africa, 1899–1902, he was on terms of friendship with Milner and members of the Round Table (see note 22). Like them, he was devoted to the idea of empire. MP for Birmingham South, 1911–1945; he served in Intelligence 1914–16, and then in the War Cabinet. In 1919 he became parliamentary under-secretary to Milner at the Colonial Office, moving to the Admiralty in 1921, First Lord, 1922–3. Colonial Secretary, 1924–9 (and of the Dominions, 1925–9), he instituted a development programme for the colonies: he established the Empire Marketing Board, and a colonial agricultural research service. He was secretary of state in the India Office, 1940–5, at odds with Churchill over self-government. It was said of him that he might have been Prime Minister – if he had been six inches taller and his speeches thirty minutes shorter. He demolished Asquith in 1916 ('For twenty years he has held a season ticket on the line of least resistance, and gone wherever the train of events has taken him') and Chamberlain in 1940, quoting Cromwell: 'You have sat here too long for any good you have been doing – in the name of God, go', in the debate on the eve of Chamberlain's resignation.

20. Philip Henry Kerr, 11th Marquess of Lothian (1882–1940). Robert Brand (banker and financier; Fellow of All Souls 1901–1932, 1937–63) recruited him to his own staff in South Africa in 1905: he became a member of Milner's Kindergarten, responsible under Brand for railways in the four states. He was the first editor of *The Round Table*, in 1910, and became Lloyd George's private secretary, 1916–21. He was Secretary to the Rhodes Trust 1925–39. As Marquis of Lothian from 1930 he became Chancellor of the Duchy of Lancaster and then Ambassador to Washington in 1938 until his death in 1940 allowed Churchill to place Halifax, most successfully, as ambassador during crucial war years. Kerr was a convert to Christian Science in 1925 and his early death is attributed to his refusal to accept

conventional medical treatment. Contemporaries thought him impressionable and superficial. He had been a candidate in All Souls in 1904, but was not elected.

21. J.L. Brierly (1881–1955). Charterhouse and Brasenose. He had been a Prize Fellow of All Souls 1906–13. He became Chichele Professor of Private International Law (and again Fellow of All Souls) 1922–47. In 1933–5 he was a member of Hebdomadal Council. He became Montague Burton Professor of International Law, Edinburgh, 1947–51; UN law commissioner, 1948–51. He was an enthusiastic supporter of Beveridge's Academic Assistance Committee, helping to place German Jewish refugees in British Universities in the later 1930s.

22. Lionel Curtis (1872–1955). Haileybury and New College. Fellow of All Souls 1921–55. He was a multifaceted fixer-supreme, a man of influence with those who held high office. He had shown great administrative ability serving under Milner in South Africa after the second African War, becoming in some sense the lynchpin of 'Milner's Kindergarten' – of which more below. Concerned with economic and social and political reconstruction, the Kinder achieved what they achieved, but certainly acquired a lasting sense of the meliorative potential of the British Empire. The Round Table was a group of ex-Kinder, Curtis its Secretary, as well as a journal of idealistic imperialism, Kerr its first editor. He was Beit Lecturer in Colonial History in Oxford 1912–13, and secured Coupland's succession to the Chair held by H.E. Egerton in 1920. He was the inspirer and organiser of the Institute of International Affairs in 1919: it later became the Royal Institute, established in Chatham House. In 1935 he published the first of the three volumes of his *Civitas Dei*, devoted to the theme that an Empire, not necessarily British but properly organised and run by the right people, would indeed approach a heavenly state in this world (Lavin 1995).

23. Thomas Penniman got a distinction in the Diploma in 1928; but he then seems to have lived from odd scraps of money, for instance £50 for the Secretaryship of the Committee for Anthropology from 1933, and to have had no established position that appears in the *University Calendar*. He became Curator of the Pitt Rivers Museum in 1939. He wrote mostly about items in the Pitt Rivers collections, with a special interest in metallurgy; he was an occasional junior author with Marett, and a collaborator with Beatrice Blackwood.

24. W. D. Ross, FBA, a classicist.

25. Hugh Edward Egerton (1855–1927). Rugby and Corpus. He was a lawyer and secretary to Edward Stanhope, his cousin, a Fellow of All Souls (1863–70) and Secretary of State for the Colonies 1886–7 under Salisbury (also a Fellow of All Souls, 1850–3). From 1905 Egerton was Beit Professor of Commonwealth History with a Fellowship at All Souls (1906–20). His work was admired in his time. Lionel Curtis was the Beit Lecturer in 1911–12, and Alex May (in *ODNB*) records Egerton as saying that it was like being a country rector, with the Prophet Isaiah as curate.

26. Charles Grant Robertson (1869–1948). Highgate and Hertford, Fellow of All Souls 1893–1948. He was Principal, later Vice-Chancellor of Birmingham, 1920–38, where he was a notable innovator.

27. Donald Somervell (1889–1960). Harrow and Magdalen, Fellow of All Souls 1912–45. He was a barrister, judge and MP for Crewe 1931–45; Solicitor General 1933–6, Attorney General 1936–40; Home Secretary 1945; Lord of Appeal 1946–54.

28. Richard Pares (1902–58). Winchester and Balliol, Fellow of All Souls 1924–45, 1954–8. Historian of the West Indies as well as of early nineteenth century England. Professor of History, Edinburgh 1945–54. He was a prodigious scholar, not at all conservative in his views, and attracted the admiration of the younger Fellows at that time. His adhesion to the motion was a significant coup. Berlin called him 'The best and most admirable man I have ever known', an opinion echoed by Rowse (Berlin 1980: 95; Rowse 1962).

29. Harry Hodson (1906–99). Gresham's and Balliol, Fellow of All Souls 1928–35. He
     was the son of T.C. Hodson, Professor of Anthropology at Cambridge. Journalist
     and professional protégé: editor, *The Round Table*, 1934–9, 1946; among other
     things, constitutional adviser to the Viceroy of India 1941–2. Editor, *The Sunday
     Times*, 1950–61; first Director of Ditchley Park, 1961–71.

30. Arthur Salter (1881–1975). Oxford City and Brasenose. He was a scion of the
     Oxford boat-building family. He joined the civil service in 1904, and from 1911
     worked on Lloyd George's national insurance scheme. He did vital work on the
     organisation of shipping during the war, and (with the League of Nations) on the
     financial and economic reconstruction of Europe. He became Gladstone Professor
     of Government and administration in 1934, with a Fellowship at All Souls,
     1934–75. MP for Oxford University 1937–50.

31. It is difficult to know what kind of interest this might refer to: then as now, people
     of a Hampstead and wiseacre disposition found snippets of the discipline helpful in
     reinforcing their common sense of the fundamentals of human nature or human
     sociability. Rather few such 'interested' people had any interdisciplinary learning.
     It is true that Marett was a classicist by origin, as was Coupland. Collingwood was
     notably absent from the group that urged an anthropological appointment in the
     University. J.L. Myres, Wykeham Professor of Ancient History 1910–39, was an
     active member of the Committee for Anthropology, had been Hon. Secretary and
     President of the RAI, and certainly justifies the claim so far as archaeology was
     concerned. It does not seem that Frazer had a following among Oxford classicists,
     as he did in Jane Harrison and others in Cambridge. History had to wait another
     twenty-five years for Keith Thomas, and Oxford Theology still awaits its messenger.

32. Geoffrey Faber (1899–1961). Rugby and Christ Church, Fellow of All Souls
     1919–61. Estates Bursar at this time. Poet and publisher. He served in the Post
     Office Rifles 1914–19. He had intended to join the ICS, but went into partnership
     with Maurice Gwyer (see note 48) and his wife to run Gwyer's wife's inherited
     publishing business. From 1929 this became Faber and Faber. He married Enid
     Eleanor, daughter of Sir Henry Erle Richards, Chichele Professor of International
     Law and Diplomacy, and Fellow of All Souls 1911–22.

33. Ernest Woodward (1890–1971). Merchant Taylor's and Corpus, Fellow of All
     Souls 1919–71. Domestic Bursar at this time. Professor of International Relations
     (Balliol) 1944–7; Professor of Modern History (Worcester College) 1947–51;
     Professor, IAS (Princeton) 1951–61. Served in Artillery and Intelligence
     1914–18; Intelligence and Foreign Office 1939–45.

34. The College had a more practical interest in psychology: the Academic Assistance
     Council (formed to find support for Jewish refugee scientists from Germany) had
     recommended the case of Dr Erwin Strauss to the College, which was supportive.

35. Berlin urged great caution in decisions about a discipline 'on whose fringes so
     many curious fanatics, eager but confused dilettanti, and occasionally out-and-out
     charlatans are still to be found' (Berlin 2001).

36. Faber to Berlin, 2 October 1936: 'the kind of psychologist I want to encourage is the
     man with a philosophical background and training ... to make sense of the
     empirical advances made by the psychotherapists'-a want rather far from Berlin's
     interests. Similarly H.H. Price, at that time Wykeham Professor of Logic and a
     Fellow of New College, who commented on Berlin's paper 'I am sorry you found no
     room for Psychical Research ...'. These letters, brought to my attention by Henry
     Hardy, are in Bodley, Special Collections and Western MSS, MS Berlin 105, 88–9
     (Faber), 80–82a (Price).

37. This committee was established precisely to allocate surplus money to research: not
     to other uses. Its membership in March 1936 was: Adams (Warden), de Zulueta
     (Sub-Warden), Oman, Robertson, Archibald, Woodward (Domestic Bursar), Faber

(Estates Bursar), *Coupland, Macgregor, *Brierly, *Pares, Rowse, *Hodson, Sparrow, Clark, *Salter. The names marked with an asterisk are men who signed the original motion to support the chair of Social Anthropology. They are six out of sixteen, but I cannot find the committee minutes at present, and do not know who was present at the crucial meeting, nor whether they took a vote on the issue.

38. The University raised the CUF fund on a College's net income, after various allowances which until 1967 included professorships, readerships, lecturerships and contributions to Bodley. Bailey points out, the cost to the College (provided the total of these expenses was less than the tax to which the College would otherwise have been liable) was nil. Research and other sorts of Fellowship, however, were not tax-deductible, and had to be paid for from taxed income. To choose tax-deductible posts gave the College some control over the direction of academic development in the University, which it would not have had by simply paying tax (Bailey, nd: 11).

39. John Simon (1873–1954). Fettes and Wadham, Fellow of All Souls 1897–1954; interalia Solicitor-General 1910; Attorney General 1913; Home Secretary 1915–16; Foreign Secretary 1935–7; Chancellor of the Exchequer 1937–40; Lord Chancellor 1940–5. He was a member of the Asquith Commission. His Opinion was never tested in a court, but was circulated to Hebdomadal Council in 1909 and 1913: it was referred to in later meetings.

40. Those who are interested might note the areas in the London A – Z colonised by All Souls names: page 26, near Hendon Station, has roads named after Faber, Woodward, Edgeworth, Talbot, Malcolm (including Malcolm Park). Page 58, near Kensal Green station, has All Souls Avenue, with tributaries called after Fellows: Bathurst, Buchanan, Holland, Leigh, Leighton, Amery, Clifford, Doyle, Egerton, and Chelmsford.

41. This paper is about academic activity, rather than about investment. However, do note that the College moved out of ground rents from about 1935: with the increasing probability of war, the Bursar and Warden of a College with a reputation for appeasement thought it prudent to protect the College from war inflation (Bailey, nd). Bursar Faber and Warden Adams persuaded the College to invest in agricultural land, buying when the yield would be about 4 per cent. Bailey notes that leases on urban land sold at between £700 – 1,000 per acre; while the College bought freehold agricultural land as a hedge against wartime inflation at £15–20 per acre. The College's agricultural estates doubled between 1935–42, from about 10,000 acres to about 21,000. People sometimes imagine that College endowment is a passive element in the College's wealth. But one of the lessons of All Souls' history in the period 1880–1940 is that it requires accurate strategic planning, fixity of purpose and energetic pursuit of agreed aims.

42. These were: 1905, Beit Professor of Colonial History; 1909, Chichele, of Military History; 1912, Gladstone, of Government and administration; 1919, Marshal Foch, of French Literature; 1931, Chichele, of Economic History; 1936, Spalding, of Eastern Religion and Philosophy; 1936, Social Anthropology.

43. Curzon (1899–1905); Chelmsford (1921–6); Wood, later Halifax, (1926–31).

44. Born in Mysore, father in the ICS. Fellow 1894–1994.

45. Born in Mysore, son of the Chief Justice.

46. Fellow 1932–9, 1969–99. Son of T.C. Hodson, ICS 1894–1901, who became Reader in Ethnology in 1926, and then William Wyse Professor of Anthropology at Cambridge, 1932–7.

47. Fellow 1889–97. He had himself served in the Indian Education Service 1892–1919, and was a frequent visitor to College after his retirement, though it seems he became reclusive after 1931.

48. Fellow 1902–16. He was the chief architect of the Government of India Act of 1935, and became Chief Justice of Federal Court of India in 1936.

49. Born in Bangalore 1868. Fellow 1925–51. He had served as a military engineer in India 1890–96, and in South Africa 1898–1906. He became chief instructor in fortification at Woolwich in 1906, and then served in the historical division of the Committee for Imperial Defence until 1925, when he became Chichele Professor of the History of War, with a Fellowship at All Souls.

50. Fellow 1914–21. He was Professor of Modern History at Allahabad, and worked with Lionel Curtis on Indian constitutional reforms. He also served as 'foreign minister' to the Maharaja of Patiala and (in that capacity) attended the League of Nations in 1925.

51. Penderell Moon (1905–87). Winchester and New College, Fellow of All Souls 1927–35 and 1965–72. He entered the Indian Civil Service in 1929, was dismissed in 1943 and re-instated in 1946. He remained in India after independence, as Chief Commissioner in Himachal Pradesh and of Manipur, retiring in 1961. Knighted 1962.

52. Geoffrey Robinson (Dawson from 1917) (1874–1944). Eton and Magdalen, Fellow of All Souls 1898–1944. On Milner's staff 1901–05; editor Johannesburg Star 1905–10; The Times 1912–19, 1923–42; Secretary, Rhodes Trust 1921–2; Director, Consolidated Gold Fields of South Africa, etc. With Milner and Amery conspired the fall of Asquith in 1916. In the 1930s he was a leading appeaser.

53. Robert Brand (1878–1963). Marlborough and New College, Fellow of All Souls 1901–32, 1937–63. He was on Milner's staff 1902–09, and a director of Lazard Brothers 1909–60. He was a member of the editorial board of the Round Table, served on the Imperial Munitions Board (Ottawa) in 1915, and at the Paris peace conference 1918–20. In the Second World War he headed the British food mission in Washington, and was the chief Treasury representative in Washington 1944–6.

54. Dougal Orme Malcolm (1877–1955). Eton and New College, Fellow of All Souls 1899–1915, 1921–55. He worked on Milner's London staff from 1899, becoming Private Secretary to Lord Selborne (Milner's successor as High Commissioner).

55. Dr Nevill Colclough has said that to adopt a discipline because of the quality of insights it afforded was like buying a washing machine because of the quality of the free plastic bucket that comes with it.

56. When I entered All Souls in 1990, I was greeted by Isaiah Berlin (who played his part in the decisions of 1935–6): 'I have known all your predecessors: two charlatans, one eccentric and one sensible man. I wonder what you will turn out to be.'

57. So far as I can tell, he was egging the pudding: the College's policy was unchanged since Warden Adams's letter of 1 June 1936.

58. It is not at all certain that R-B was popular in Faculty or College; certainly Beatrice Blackwood thought poorly of him (see Mills, this book). On 7 February 1946 R-B wrote a further letter to Sumner stating the case for retaining the 'social' qualifier, and enclosing a paper he had written for *Nature* on 'The meaning and scope of Social Anthropology'. Since he had written this forty-two years previously it might be thought a little off-hand to give it to the Warden as a guide to the discipline or an introduction to his subject. That would fit with his surviving reputation in All Souls. The papers are in ASC, MS 435.

59. A.L. Rowse chaired the meeting as acting Warden: Sumner died in early 1951.

60. This was financed by the UGC. It is worth noting that, while the salaries of University post-holders increased, especially after the war, that extra cost was borne entirely by the University: the various Colleges' share was a diminishing proportion of the total salary bill.

Chapter 4

A MAJOR DISASTER TO ANTHROPOLOGY?

OXFORD AND ALFRED REGINALD RADCLIFFE-BROWN

David Mills

Introduction

Alfred Reginald Radcliffe-Brown liked to don a monocle, a cloak and more than his fair share of Edwardian manners. In that tactful tone required of contributors to the *ODNB*, Godfrey Lienhardt describes him as 'somewhat aloof' and 'politely overbearing'. His haughtiness and emotional distance won him more enemies than friends. The title of my chapter is taken from a letter written by Beatrice Blackwood, long-time teacher of ethnology at the Pitt Rivers Museum. She described how she and Penniman – curator of the Museum after Henry Balfour's death in April 1939 – were 'fighting to keep the present Diploma' in the face of a 'sneeringly obnoxious' R-B. For her, R-B was a 'major disaster to anthropology in Oxford ... losing for us all the ground we have gained for anthropology in Oxford for the last forty years'. She bemoaned the death of Leonard Buxton as a 'sad weakening of our forces, as he could have stood up to RB, and could turn on a tongue as cutting as RB's own' (PRM Archives, Blackwood papers, Blackwood to Skinner, 16 November 1939).

Was R-B really a major disaster to anthropology? In this chapter I explore his attempts at reforming anthropological research and teaching at Oxford. I dwell less on his disputed intellectual legacy, a topic on which much ink has already been spilt, than on his bureaucratic skirmishes. These include an attempt to introduce an undergraduate Final Honour School, a battle to reshape the broad one-year postgraduate diploma into three more specialised diplomas of which one would be social anthropology, and his effort to shift the department's focus towards the supervision of graduate research students. I discuss each in turn. I suggest that the changes Malinowski wrought at the LSE provided a model for R-B's own reforms. His

Figure 7. Alfred Reginald Radcliffe-Brown, Professor of Social Anthropology 1936–46. Copyright ISCA, University of Oxford.

interventions, whilst clumsy and often badly handled, did promote the interests of his vision for Oxford anthropology, albeit at the expense of the two other branches of the discipline. His Presidency of the RAI saw him initiating similar reforms, a process that eventually led to the foundation of the Association of Social Anthropologists (ASA) in 1946. Whilst R-B's imperious and doctrinaire ways were disliked, the legacy of his reforms for the fortunes of social anthropology were anything but disastrous.

Anthropology as a Final Honour School?

Radcliffe-Brown was appointed to the Oxford professorship in July 1936. Few of the papers relating to the election remain, though there were more than a dozen candidates, including, by his own account, Evans-Pritchard. One of the members of the appointment panel, the historian Reginald Coupland, originally invited Malinowski to apply for the post. Malinowski ruled himself out of the race, explaining that he had 'a very real debt of gratitude towards the School of Economics and the University of London' and 'that I would not like to sever my connection with this institution which has assisted me so generously and effectively in developing social anthropology in this country'. So instead he proposed R-B as 'by far the most suitable from every point of view', proclaiming his 'genius' at organising departments as much as his theoretical contributions. Coupland responded by expressing disappointment that 'they must be content with second best', but expressing the hope that R-B would indeed actually apply for the post, as 'the electors might not think R-B's claims so outstanding as to sit all our candidates aside and issue an invitation which might be refused' (British Library for Political and Economic Science (hereafter BLPES), Malinowski archives: Malinowski to Coupland, 2 July 1936; Coupland to Malinowski, 31 July 1936).[1]

Much has been made of the public rivalry and seeming enmity between the two champions of the new school of anthropology. R-B's letters show a different dimension to this relationship, as he repeatedly sought to flatter Malinowski, asking him for advice and support, and even sharing confidences about his chest problems. Back in 1929, whilst in Sydney, R-B had confessed to Malinowski that 'the trouble about Anthropology is its name'. He went on to bemoan the inclusion of physical anthropology and prehistoric archaeology in a 'conventional curriculum', and wished 'never to have to teach either physical anthropology nor archaeology again'. In a moment of candour, he proposed that 'you and I and anybody else who will help us ought to build up the new sociology or anthropology that is needed', preferably in Oxford, because 'Cambridge does not suit my health'. He even felt Oxford was a risk for his chest, but a risk that would be worth it 'if I was free to treat the subject in my own way and not be required to lecture on the 'Races of Man' etc'. The letter ended with a further bout of flattery – 'I should greatly like to be working near you and co-operating more closely'. So it came to pass. Whatever Malinowski thought of his protagonist, his intervention was influential, and R-B was, on paper at least, grateful (BLPES, Malinowksi archive: R-B to Malinowski, 3 December 1929).

Whilst Malinowski had left LSE for a sabbatical in the USA in 1938, his success in getting his particular version of anthropology institutionalised within the LSE in the early 1930s would have been closely watched by R-B. Malinowski's vision of postgraduate research-led teaching had partly been made possible by his close relationship with Jessica Mair, Lucy Mair's aunt and the personal secretary of Lord Beveridge, the LSE's director. Along with fund-raising, academic politicking was Malinowski's forte, and he gradually marginalised Professor Seligman with his high-profile research seminar and cohort of research students. Such far-reaching pedagogic reforms were less simple within the Oxford system.

Radcliffe-Brown finally arrived in Oxford in October 1937. The delay, given that he had been elected in July 1936, left Marett to carry on the duties of the new professor, including haggling with the geographers over space for the expanding department within their new shared premises on the corner of Mansfield Road and Jowett Walk. During the summer of 1937 Marett dispatched R-B a series of curt notes on his departmental responsibilities, such as paying the caretaker! Obviously keen to meet, Marett invited him to come and visit him in Jersey where he was 'busy digging out a cave' (Exeter College, Oxford, Marett papers, L–IV–11).

Despite his late arrival, R-B had been doing groundwork of his own, and had already begun to develop his vision for a more specialised research-led field. Flattering the influential classicist John Myres after his appointment, he wrote to express his appreciation at being 'called to become your junior colleague in the oldest school of anthropology in the British Empire', and seeking his advice over the 'further development of social anthropology at Oxford' (Bodleian, Myres MSS 81: 25).

As others had done before him, R-B was attracted to the idea of a FHS in anthropology, with candidates being 'permitted to specialise to a limited extent' within one of the four fields. As a result, in January 1937 he wrote to his colleagues about his ideas, proposing a series of fifteen written exams, with five compulsory and four optional papers. His suggestion that 'candidates specialising in physical anthropology or prehistoric archaeology might be expected to complete their work in one year but candidates in physical anthropology and prehistoric archaeology or social anthropology should be strongly recommended to take two years for their work'.

But any proposal he drew up had to be vetted by the Committee for Anthropology, which along with Marett numbered Henry Balfour, Leonard Buxton and Le Gros Clark. At first, they were in agreement, and in his absence, a sub-committee prepared a fuller draft for the degree, which included a comparison with the London and Cambridge degrees, and the numbers of students they attracted. But they also put their own spin on the proposal, recommending that an FHS be based 'on a combined study of Race, Culture and Evolution in their combined bearing on the evolution of society'. The committee took the view that the combination 'has hitherto worked so well that there is no reason to depart from it'. The sub-committee concluded that the FHS should retain the principle of the present Diploma and 'be on a broad educational basis'. It was decided to await the reply of the new Professor before submitting the new statute (OUA, DC 1/1/2: 235, March 1937).

This was hardly what R-B had in mind. But after arriving in Oxford in October 1937, he was unable to bring the Committee around to his vision for the degree or for reforming the discipline. Whilst they were all for a FHS, they each had rather different views about its contents. For his part, at the first meeting R-B also proposed reforming the Diploma, as he realised the slow time-scale for the introduction of an Honour degree. He was also suspicious that the University would not adequately provide for the extensive teaching load accompanying an undergraduate programme.

The institutional constraints on introducing a FHS became clearer at a meeting he held with the University Registrar in March 1939 to seek support for his vision for the department. The Registrar, Douglas Veale, in a subsequent briefing note to the Vice-Chancellor, reported that the Professor had announced that 'unless he could develop a school of anthropology, preferably an undergraduate school ... a real cultural school, he would much rather go away'. He named his price – £600 a year. Veale retorted that the only way to make such a sum was to save it elsewhere, asking whether 'there was any need for a reader in Physical Anthropology at £550 a year'. R-B agreed wholeheartedly, saying that 'people had been measuring skulls for 60 years without producing a single result of real scientific importance', and that money for the development of social anthropology could best be obtained by retitling the Readership in Physical Anthropology. He also pointed out that Nuffield College was keen to draw on anthropological expertise in developing its Colonial Studies provision (OUA, UR 6/ANT/4: Memorandum by Veale, 18 March 1939).

Veale was not convinced, and asked about the competition for undergraduate students with Cambridge (which had 60 students at this point) and London, as the 'total number of Undergraduate students is fairly strictly limited'. He again wrote to the Vice-Chancellor that he was 'of the opinion that anthropology was unlikely to become part of an 'intellectual education'. As he explained, 'it seems to me unlikely that Anthropology will ever become a school like Greats, Modern History or English Literature, which will be taken merely as an intellectual discipline by people who intend to follow careers for which special knowledge is not required'. He went on to say that 'fifty years ago there might have been a stronger case for establishing Anthropology, but it was not done and the claim is now very much weaker'. He concluded that 'the time has arrived when we ought to concentrate on what we are doing well', leaving Cambridge to 'develop social anthropology without competition from us'. (OUA, UR 6/ANT/4: Veale to Vice-Chancellor, 10 January 1940).

These accounts may give the impression that R-B was fighting some of these battles alone. This is not the case. E-P had accompanied him on his visit to the Registrar, having been appointed to a Research Lecturership in African Sociology in 1935. In 1939 he was joined by Meyer Fortes, who also held a Research Lecturership for two years, and also by Max Gluckman, before he left to take up a post at the Rhodes Livingstone Institute in September of that year. This small coterie met regularly with R-B, both in his rooms at All Souls and at various North Oxford public houses, and over which he continued to cast an intellectual spell. As E-P later recalls, it was these discussions over 'system' and 'structure', later reiterated in the 1940 RAI Presidential address 'On Social Structure' that formed the analytical underpinnings of *African Political Systems* (Fortes and Evans-Pritchard 1940) and *African Systems of Kinship and Marriage* (Radcliffe-Brown and Forde 1950). But the relationship between 'master' and 'student' was not straightforward, given E-P's own charismatic personality, and the fact that, by some accounts, he had come a close second to R-B in the 1936 election.

Radcliffe-Brown spent a great deal of energy pushing for a one-year diploma that specialised in social anthropology. Part of his case for change was the necessity to make 'better provision for an important class of colonial officials who wish to study social anthropology, but who have not the time at their disposal for the full diploma course'. It was not the first time that anthropology's utility beyond the university was touted in order to entrench the academic discipline more firmly. No revision of the Diploma regulations had been made since 1905, despite the theoretical developments in the discipline, another point which R-B saw as playing in his favour. His proposal was to have a general one-term introductory course, allowing admission into a two-term specialisation for 'advanced work' in either social anthropology, or physical anthropology or technology, with colonial officers excused the first term 'in consideration of their previous experience' (Bodleian, Myres MSS 81: 33, 138).

The proposal, like R-B's other initiatives, was highly controversial. The General Board of the Faculties – a University-wide structure – was not inclined to register three separate diplomas, given the costs of examining and the precedents they would set for other departments to establish their own diplomas. But it was his academic colleagues on the Anthropology Committee who led the opposition.[2] Offering what amounted to a specialist diploma in social anthropology would mean that 'physical anthropology' and 'primitive technology' would no longer be compulsory, breaking up the long-respected 'trinity' of anthropological teaching at Oxford. Balfour was outraged, and was driven to write one desperate postcard to Myres, headlined 'This is an SOS', protesting at what he described as the 'extremely one-sided and narrow minded proposal now afoot'. He did not mince his words. He felt that a one term general introduction was a 'ludicrous allowance', and would turn out a lot of incompetent 'specialists', with too little general preparation, and not enough within their own field to be of great use (Bodleian, Myres MSS 81: 35, 131).

The question that divided them was simple. Would an application for three specialised diplomas weaken the case for a FHS that would bring together the different fields? Most saw R-B's preference for the former as working against the longer-term aim of an Honour School. During 1938 and 1939, letters and counter-proposals for reform of the Diploma flew backwards and forwards, a phenomenon Penniman described as a 'pamphlet war', as R-B sought to 'torpedo' the existing diploma. R-B did not mince his words – he was of the opinion that 'Diploma students are not worth teaching, they do not have the time to do even the minimum of reading'.[3]

Things came to a head when in late 1939 R-B circulated a memo stating that the standards of the Diploma students were embarrassingly low, and that 'both physical and social anthropology suffer very much from persons who know a little (or think they do) and do not know how little they know'. He felt that it should be made impossible for anyone to 'obtain a diploma in anthropology by spending three Terms in picking up a few miscellaneous and disconnected scraps of knowledge about a number of subjects which it is quite impossible to study systematically in that time' (Bodleian, Myres MSS 81).[4]

Penniman took up the debate on behalf of the Pitt Rivers Museum, tabling a counter-resolution to maintain the existing Diploma and to push for a FHS. Citing E-P's support for the Honour School, Penniman pointed out that the Diploma had always been intended as an introductory training, and that whilst it was 'impossible to eliminate entirely the type of student who will not do credit to his training, much can be done in this direction without wrecking the existing diploma'. Penniman felt that if they continued with such 'tinkering' and 'monkeying' they would be laughed at, and he was loath to accept the 'uncertain schemes of a man who always appears to change his mind halfway through any plan he sponsors' (Bodleian, Myres MSS 81: Penniman to Myres, 27 October 1939).

Eventually it was left to Myres to take up the elder statesman role, smoothing over the tensions with a set of highly complicated new regulations for the Diploma that all sides could agree on, though meeting many of R-B's demands along the way. It remained a single diploma, but allowed for increased specialisation in one of the three sub-fields. The final syllabus was not agreed until 1940, by which point the onset of war had made students scarce on the ground.

Radcliffe-Brown was unbowed. In the same provocative vein, he was behind the renaming of the Department of Social Anthropology as an Institute. After getting agreement from the Faculty Board to his proposal in June 1939, he then wrote to the Curators of the University Chest explaining that 'If social anthropology is ever to have any real importance in Oxford it will be because this becomes a centre for research.' He went on to note that 'there may be some chance of appealing for outside financial help' but that it 'would be somewhat easier to appeal for funds for an Institute than for a department'. Its implications would be that 'we intend (or at least hope) to be concerned not only with teaching but also with research' (OUA UC/FF/228/2).[5] The following April, long after R-B had physically changed the letter-plate on the door of the Institute, the General Board of the Faculties rubber-stamped the retitling.

The new title was no flight of fancy. R-B was aware of the Colonial Office's plans for a major Colonial Development and Welfare Act, and wanted to ensure Oxford was able to benefit from the research largesse that Lord Hailey – author of *An African Survey* (1938) and architect of the act – had insisted be part of such a programme. In the same year he applied to the University's Higher Studies Fund for resources to support a programme of research training in social anthropology. It was accompanied with another threat, testimony to the frustrations he faced over the FHS and the specialist diploma. He hinted that if 'the University decides to take no effective part in the development of a subject the importance of which in a colonial empire is being increasingly recognised' it would 'leave serious work in anthropology to Cambridge and London' (OUA, FA 4/2/1/1: 80).[6]

The story of anthropology's marginalisation by Nuffield College demonstrates some of the challenges R-B was facing within the university. The college was founded in 1937 with a one-million pound endowment from the car-maker William Morris. In his original proposal to the University, he envisaged a place in which 'to bridge the separation between the theoretical students of contemporary civilisation and the men responsible for carrying it out'. Whilst his intention was for an undergraduate college and a vocational engineering institute, senior figures (including the Vice-Chancellor, the Warden of All Souls and the Master of University College) in the University recast its role as a post-graduate college to support the nascent social sciences, aware of their current reliance on a series of rolling Rockefeller grants in this field (such as the one that funded E-P's research lecturership until 1940).

The research direction of the new college was to be informed by two sub-committees – Politics and Economics. Despite having one seat on the Politics committee, little provision was made for the discipline, leading R-B to write to the Research Committee in early 1939 to ask how 'how the claims of Anthropology and Geography will be met by arrangements for Nuffield College'. Later that year, he submitted a costed proposal for a new Anthropological Institute, 'with the suggestion that some or all might be found by Nuffield College'. The idea was rejected by the Research Committee, and the request led the College to define its spending principles: 'money shouldn't be expended on objects which are inconsistent with the aims of Nuffield ... or frittered away on minor objects'.

Paradoxically, the first research work carried out within the college was that done by Oxford anthropologists during the war. Margery Perham, an influential public commentator on colonial issues, was appointed University Reader in Colonial Administration and Director of Research at Nuffield in 1939. After failing to commission Isaac Schapera to look at migrant labour in Central Africa, she co-ordinated a major economic survey of Nigeria, carried out by Daryll Forde and involving Meyer Fortes and Kofi Busia, then a research student at the Institute, and funded by the Colonial Social Science Research Council. There is no record of anthropological research carried out at Nuffield after the war (Nuffield College, Committee minutes 1937–40 (Volumes 1–8).

Anthropological In-fighting: The RAI and the ASA

Radcliffe-Brown's veiled ultimatums were not just reserved for Oxford's academic politics. There was a similar jockeying for power within the RAI when R-B was elected President in 1939. For R-B, the

RAI represented a potential threat to the claims that the nascent university social anthropology departments might have over Colonial Office research funds.

The Secretary William Fagg, and the editor of *Man* Ethel J. Lindgren, both schooled in a view of anthropology that embraced its physical and biological aspects, resented R-B's attempts as RAI President to 'run down' or remould the Institute as 'mouthpiece for his particular brand of social anthropology'. They also were far from happy with the way that the RAI decamped to Oxford, where it held a special meeting in December 1939 because of the travel and black-out restrictions imposed by the war. Writing in 1940 to Herman Braunholtz, William Fagg expressed his fears over 'RB's defeatism' and the future 'ruin' of the RAI:

> Radcliffe-Brown seems to be hag-ridden with the idea that there is a New Order impending in this country in which there will be no room for the Institute; he seems bent on shutting down the Institute for the duration, maintaining perhaps its skeleton at Oxford (or even in Raglan's vaults). ... Either anthropology will be left without an effective central organisation, or a rival body (of which I have often heard dark threats) will be set up at Cambridge (RAI, House archives: 6/28).

For this reason, Fagg vowed to stay on as RAI Secretary, despite R-B's attempts to replace him with Raymond Firth. At this point, Fortes, acting as secretary of the RAI's newly revived Applied Anthropology committee, sought the support of RAI Council for a memorandum to the Colonial Office on the need for anthropological contributions to research in the colonies. Fortes and others wished to stress the potential contribution of social anthropology. The question arose whether the memo should also mention the study of material culture and technology. Fagg was far from enthusiastic about the proposal, suspicious of Fortes' position in the R-B camp, and feeling that he was not acting with the RAI's interests at heart. Writing to Fagg, Lindgren warned that 'we must, of course, be prepared to have our plan defeated by the President, who is accessible to influences from Fortes at Oxford' (RAI, House archives: 95/7/10).

The first draft called again for a central bureau, such as the RAI, to co-ordinate research. R-B rejected it, and prepared an alternative draft, titled a 'Memorandum on the Hailey Report'. R-B insisted that 'research into the anthropological side of colonial problems will yield best results ... if carried out from research institutions in this country'. He went on to argue 'that in the first place the task for research as well as that of training should be entrusted to British Universities' with 'departments of Social Anthropology headed by a Professor of the subject'. This amounted to a call for an expansion of appointments at

Oxford, Cambridge and London. Predictably the RAI officers were furious, as it made no attempt to build upon what Lindgren felt had already been achieved; namely 'anthropological training for all colonial probationers, anthropological research by government officials, the appointment of government anthropologists'. For her, it simply emphasised a 'purely academic viewpoint' (RAI, House archives: 43/13/13, 43/13/10).

At this point things became more machiavellian. Fagg nurtured Hutton's opposition to this 'new' Anthropology and Myres' growing unease with R-B's leadership at Oxford in order to mount a counter-offensive. Capitalising on what Fagg later described as a 'considerable error of judgement on Radcliffe-Brown's part' in not consulting Seligman whilst drafting his document, Fagg ensured that the final meeting to approve the memorandum was packed with people aggrieved at the manner in which R-B had handled the matter. Later writing to Lindgren, he commented with satisfaction that 'the enemy came prepared to negotiate'. The final draft ended with the compromise that whilst 'the setting up of research institutes in British universities and in the colonies themselves is recommended as the primary necessity, plans should be made for establishing an Institute of Colonial Anthropology' (RAI, House archives: 13/25/21).

As Fagg later acknowledged, comparing the event to his fears over the current war, 'the victory of which I spoke was of course purely relative ... we have neutralised the worst features of the original draft, and ensured that our own sociology Hitler will not achieve 'world domination". Fagg feared that the neutering of the memorandum, would 'hammer anthropology in general by showing us up as a muddle-headed lot who have the greatest difficulty in restricting our chauvinistic elements', adding that the Colonial Office 'may think we must put our house in order before we expect much consideration'. He was not far off the mark. The British Colonial Secretary Malcolm Macdonald confessed in 1940 that he found 'anthropologists as a class are a difficult folk to deal with' (Rhodes House, Oxford, British Empire Manuscripts: MacDonald to Hailey, 18 April 1940).

The repercussions of this feud were far-reaching. The festering division within the RAI eventually led to a schism in 1946. E-P decided to use his appointment to the Oxford chair to launch a professional association dedicated to social anthropology. It was an idea that he, Fortes and R-B had already been discussing, but E-P went on to invite Siegfried Nadel, Max Gluckman, Brenda Seligman, Raymond Firth, Daryll Forde, Edmund Leach and John Layard (along with the archaeologists Louis Leakey and A. Arkell) to attend a first meeting in Oxford in June 1946. Ten days beforehand, he circulated a short one-page memo. 'It is suggested', he began, 'that Social Anthropology is now sufficiently distinct a study to have its own association and journal and that a co-operative undertaking of the kind is desirable in the interests of the science'. Diplomatically, he went on to insist that the association should not be 'in rivalry with existing institutions', suggesting that it might even be an autonomous section of the RAI. He carefully articulated what he saw as being the objects of the association, which included 'a) to propagate the interests of Social Anthropology, particularly by strengthening the existing university teaching departments and encouraging the formation of others; b) to co-ordinate research; c) to constitute a body, representing the interests of the science as a whole, to which governments and other corporations desiring advice on questions of research can apply' (BLPES, ASA archives: 1.1). These aims were adopted word for word at the first meeting of the Association, as I show in a short history of the Association (Mills 2003b).

Everyone E-P had invited was present, and it was resolved 'that a professional association of teachers and research workers in Social Anthropology be here and now formed as an independent body'. At the first meeting, chaired by R-B, E-P was appointed 'Chairman and Secretary-General', R-B President, whilst Firth, Forde and Fortes made up the committee. The last bit of business was to draft a letter to inform the RAI of the new association, 'hoping for collaboration with the Institute'. No mention of this letter was made in the RAI council minutes, nor was there any formal response (BLPES, ASA archives: 1.1).

How did R-B's own tenure end? With the coming of war, there were few students in Oxford. R-B found himself increasingly isolated and with little to do. He put himself forward for government service, but his offer was not taken up, and he began to think about returning to Chicago. In correspondence with Lloyd Warner, he admitted that the war meant that he now had 'very little hope of carrying through my plan for a Research Institute of Social Anthropology at Oxford', and asked whether there was any chance of a visiting professorship, admitting to missing 'the warm friendliness and congenial company' of Chicago, where he had been a Visiting Professor from 1931 to 1937. Two years later he wrote to tell Lloyd Warner of his plans to take up a Visiting Professorship at São Paulo, as the Rockefeller foundation was 'interested in the new department of social anthropology' there. But he also admitted that 'I am much happier with Americans like yourself and some others than I am with Englishmen'. R-B arrived in São Paulo on the 6 April 1942, but continued to hanker after a way back to the United States (Radcliffe-Brown 1985).[7]

Daryll Forde stood in for R-B at Oxford during this period, becoming involved in an economic survey of Nigeria. He also prepared a memo on the future needs of the Institute, emphasising, as had R-B, the value of 'sociological investigations for all aspects of colonial development' (OUA, FA 4/2/2/1: 144).

Radcliffe-Brown returned to Oxford in 1944, only to find that his request to extend his contract beyond the formal age for retirement had been turned down. He retired to his house in the hills of North Wales in 1946 in order to write, but he could not afford to employ a servant, despite a special All Souls stipend. So he decided to take another job, this time setting up a department of social science at the new university in Alexandria. His doctors approved, feeling that it would be good for his tuberculosis-ridden lungs. After two years in Alexandria, from 1947 to 1949, he held a similar post in Grahamstown from 1951 to 1954. Throughout this time he continued to write, and to attend ASA meetings in his capacity as President wherever possible, though on one occasion in 1954 he sent his apologies, not wishing to be 'the skeleton at the feast' (BLPES, ASA archives: 1.1). After being hospitalised following a bad fall and a punctured lung, he was greeted with applause at the January 1955 meeting, and the Chair expressed the hope that he would preside over many more. It was to be one of his last, and he died in a London hospital the following October. His funeral was attended by all the ASA committee and many of its members.

Conclusion

This period in anthropology's history mirrors an important shift within the intellectual culture of British universities. With growing state largesse in the first half of the twentieth century, academics no longer saw their prime concern as being the provision of a 'character-building' undergraduate education, but rather with developing a more professionalised research culture. At an organisational level, the change required new centralised financial and organisational nous – the universities needed to be able to apply for and manage large government and philanthropic grants for research. At a personal level, the shift was often achieved by dominant, dare one say egocentric, personalities who were prepared to act as missionaries for their particular view of the world, trampling over the strongly-held views of others. This style of leadership did not fit well with the collectivist decision-making culture of the Oxford faculty boards.

A comparison of the LSE and Oxford reveals the powerful role that universities play in shaping disciplinary fortunes, determining what can be taught, when and how. In Oxford, the consensual nature of committees and the faculty boards served both to protect scholarly traditions but also to constrain change and innovation. The smaller, more personalised and informal nature of decision-making in the LSE, along with the greater autonomy offered to individuals and departments, made Malinowski's iconoclastic and research-led approach to teaching and training more likely to succeed there. Yet LSE's success came at the expense of an attention to undergraduate teaching. Oxford's vibrant post-war intellectual atmosphere may have been less captivating if the Anthropology FHS had been approved, with all the undergraduate teaching this would have necessitated.

This chapter has highlighted the vulnerability of anthropology during the pre-war period, in terms of its intellectual direction, and institutional status within British universities. For Edmund Leach 'differences of social class played a critical role in what happened in British anthropology during the first 40 years of this century' (1984: 2), suggesting that the failure of Rivers and Haddon to establish anthropology at Cambridge was partly because they were seen as outsiders to the academic class cultures of Oxbridge. Ranulph Marett was the opposite. He was a consummate operator and skilled in Oxbridge institutional politics, if rather engrained in earlier traditions of anthropological scholarship. Leach postulates that 'the greater the success of LSE anthropology, the less likely it became that the conservative establishment in Oxford and Cambridge would touch the subject with a bargepole' (1984: 7). But whilst the Oxbridge 'establishment' was able to resist for a while, it could not compete with the growing move towards a professorial, research-led academia, and the funds and international prestige that accompanied it.

Perhaps only an egocentric figure like R-B was able to challenge Oxford's unspoken social codes. Because of his hauteur, he was seen as an 'arriviste', an outsider who did not understand Oxford ways. He was treated accordingly, and his antagonistic manner provoked resentment amongst his colleagues. On the other hand, the post-war Institute inherited by E-P owed much to R-B's efforts at reform. E-P did little to acknowledge this, subtly distancing himself from R-B's divisive reputation and intellectual rigidity. Nonetheless, the centre of the British social anthropological universe shifted back after the war from the LSE to Oxford, where it remained for two decades to come. R-B may have been difficult to work with, but his legacy for Oxford social anthropology was far from disastrous.

Notes

1.   Marcel Mauss was one of R-B's referees for the Chair. Whilst Mauss had previously complained about Malinowski's 'despotism' in his letters to R-B, he was less than totally fulsome in his praise of R-B's theoretical achievements in his reference (see Mauss 1994).

2.   In November 1937, the Committee for Anthropology, probably at R-B's instigation, had requested the General Board of the Faculties to create a Board of Studies for Anthropology so that it could deal with its research students' applications. The General Board responded the following year by abolishing the Committee and creating a new Faculty of Anthropology and Geography with its accompanying Faculty Board. This was a rare occasion when R-B was granted by the university authorities more than he asked for.

3.   Penniman to Myres, 27 October 1939, this is one of a series of letters in Myres MSS 81: fols 33–44.

4.   A Memorandum, entitled 'The diploma in Anthropology', from R-B to the Anthropology and Geography Board, dated October 1939.

5.   The Faculty Board of Anthropology and Geography agreed to the change of name on 8 June 1939. R-B wrote to the Curators of the University Chest on 26 June 1939 which gave their agreement, subject to the agreement of the General Board of the Faculties on 19 April 1940.

6.   The bid was unsuccessful.

7.   Letters from R-B to Lloyd Warner, 7 October 1939 and 17 February 1940, Reprinted in the *History of Anthropology Newsletter* 12/2 (1985).

Chapter 5

'A FEELING FOR FORM AND PATTERN, AND A TOUCH OF GENIUS':

E-P'S VISION AND THE INSTITUTE 1946–70

Wendy James

Elements of myth have grown up around the success of anthropology, specifically social anthropology, in the post-War years at Oxford under E-P's guidance. Both oral memory and written accounts have tended to emphasise colourful personalities, the cut and thrust of bright ideas, the contemporary euphoria. As yet we have no film or TV series, but there is a novel – Dan Davin's *Brides of Price* (1972) portrays the Oxford anthropology circle of this period, admittedly a little disguised and larger than life (and echoes of its characters can be found in Iris Murdoch). In this chapter I try to anchor the myth, or in other words to give some 'explanation' of the special atmosphere of the time – the ethos of commitment, being part of a crusade – by sketching in the less well-known background of institutional development.[1] This story reveals the precariousness of the subject at Oxford and the often unexpected factors that helped it make its mark.

The success of social anthropology was arguably achieved despite the administrative and political obstacles it faced in the University hierarchy of committees. Because of official scepticism about general anthropology as an undergraduate subject, social anthropology took off as an autonomous specialism at the level of postgraduate degrees and research (then a very small part of the University's activities as a whole). The resulting diversity of recruitment from outside Oxford, the surprisingly international character of staff and students, and the interdisciplinary mix of their previous studies, turned out to be a tremendous strength. The pre-War and wartime researches of E-P himself, in their considerable complexity, sophistication, and variety, proved an inspirational resource for people from many backgrounds. Because his work came to be deployed in interdisciplinary ways, social anthropology was able to establish itself on a wider intellectual scene than would have been the case if a more orthodox undergraduate anthropology department or degree had been set up in the late 1940s.

The inspirational figure of E-P and his shaping of the Institute of Social Anthropology have already been documented by several people – some commentators have been participants, some observers, and some both (among whom were some of the key myth-makers). Jack Goody's detailed sketch (1995) of the first few years immediately after 1946 describes the Institute as a 'power-house' of intense work and the coming and going of strong personalities (Meyer Fortes and Max Gluckman stand out) who had in several cases known each other since well before the War. John Burton's commentary (1992) emphasises the way this period distilled some special qualities of E-P's impact on anthropology; and Godfrey Lienhardt's memorial tribute (1974) both to E-P's sharp ideas, on the one hand, and his personal capacity equally to inspire or infuriate on the other, helped confirm the mystique.[2] Mary Douglas's intellectual biography of E-P (1980), an act of homage from an inspired and inspiring disciple, is a brilliant piece of evocation and re-creation. As Richard Fardon has recently shown, her own work has ranged far and wide, productively and provocatively, away from the 'mainstream' of social anthropology – whatever that mainstream might be – but her distinctiveness as an anthropologist emerged from what 'was briefly the most significant anthropological institution in the world', the Institute as rejuvenated with E-P's appointment in 1946.[3] However, 'part of the E-P myth in British anthropology' rested on what he had previously achieved in research and his extraordinarily successful publications (Fardon 1999: 24–36, quotations at 32). The making of these classics of ethnography had depended on methodical observation, but had also engaged the creative scholarly and personal imagination of the anthropologist. As E-P was later to write, not without ironic self-reference, 'The native society has to be in the anthropologist himself and not merely in his notebooks ... The work of the anthropologist is not photographic. He has to decide what is significant in what he observes ... For this he must have, in addition to a wide knowledge of anthropology, a feeling for form and pattern, and a touch of genius' (1951a: 82).

There were important differences between the kind of research, and analysis, that E-P carried out in the first part of his career and the work of his contemporaries. The conditions of E-P's own pre-War fieldwork were never those of the peaceful island communities where Malinowski and R-B, or even Firth, had forged their anthropological careers. Nor were they those of the close and orderly district administration one might have found in the settled British colonies of the time, or certainly in those of the later colonial period. Conditions in the southern Sudan of the 1920s–30s (a really vast territory, and under the FO rather than the Colonial Office) were more those of the open, often violent, 'frontier' period of primary colonisation. All E-P's field researches took place against a large stage-setting of regional variety and political process. The Anglo – Egyptian government in the

Figure 8. Edward Evans-Pritchard, Professor of Social Anthropology 1946–70. Copyright All Souls College, Oxford.

Sudan still had somewhat uncertain relations even with the Zande kingdoms when E-P first arrived, and had only just 'pacified' the Nuer by punitive military action when the government required him to study these people (Johnson 1982, 2007). The context of E-P's 1940s work on the border with Ethiopia, and then in Cyrenaica as part of the British military administration, was directly war-affected. The imperial context of all these fieldwork experiences was arguably more of an eighteenth or nineteenth, than a mid-twentieth century, character – as he himself is known to have boasted (Goody 1995: 65). The 'transparency' of his accounts, in my view, amounted to far more than a robust ethnographic writing style, despite Geertz's witty remarks (1988). In so far as models drawn from E-P's own analyses of regional or kingdom-wide political relations were later applied on a more domestic scale to relatively peaceful circumstances in the well-ordered colonies of the 1950s, they were obviously open to later accusations of 'functionalism'. But the example of his own fieldwork, of a scope not matched by anything else that had been done in Oxford, led quite easily into E-P's post-War appropriation of history (rather than natural science) as anthropology's twin. The appearance of some of R-B's much earlier essays under the banner of 'structure and function' in 1952 had very little to do with the kind of anthropology actually being fostered in Oxford by this time.

Anthropology in Oxford as elsewhere was profoundly affected by the Second World War, and its reassembly afterwards was shaped by the experiences of all during that time. E-P himself had explicitly taken on historical research in Libya, and been drawn to Islamic mysticism, before being converted to the Catholic faith in 1942; Franz Steiner arrived as a refugee from Nazi Germany and witness to the holocaust; R-B had found little to do in the early 1940s, before departing for São Paulo, leaving Daryll Forde as caretaker of the Institute and of the project to establish anthropology as a social science (though Steiner designated himself keeper of the Oxford school). Shortly after the time of R-B's return, many others, both dons and students, were returning to Oxford from the theatres of war, and there is no doubt that a feeling developed across the University that a new start was required. For anthropology, the new start was the appointment of E-P.

Proposal for an Honour School of Anthropology, 1948–9

When he first took up the Chair in 1946, E-P worked very hard to establish anthropology on a broad, inclusive basis. After all, his own research work had already spanned many strands of anthropology – he had photographed heads in the Sudan for their shape, collected objects for the museum, made drawings and musical recordings, and collected folk tales and poetry. Looking back at the detailed documents, the proposal for an Honour School of Anthropology which he co-ordinated over several years was ahead of its time, for as it turned out Oxford was not yet ready for such a degree. The 1948–9 plans, even down to some of the details, were remarkably in harmony with the degree in Human Sciences agreed in the late 1960s (admitting its first students in 1970). There were to be three main branches of study: Social Anthropology, Archaeology and Ethnology, and Physical Anthropology including Human Biology. Special subjects were to be examined, chosen from a list of eight across the board. Finally there were to be two papers in Regional Studies, chosen from another list of eight, and treated in a very scholarly way – the paper on Islam and Arab Asia, for example, included elementary Arabic, and that on the British Isles from early times to the Roman era required a knowledge of Latin. The introductory justification claimed that 'recent developments in the applied natural sciences lend emphasis to the need for integral studies of man, which related his biological characteristics to his unique quality as the creator and transmitter of culture' (OUA, UR 6/ANT/4: Proposed new Honour School of Anthropology). The proposal suggested that such studies should be recognised as a 'vital aspect of modern University education in general'. There was also a reference to the urgent need to study the problems of 'social and cultural adaptation ... among the less developed peoples of the Empire'.

The Anthropology and Geography Faculty Board endorsed the proposal in May of 1948, but questions were raised by the General Board's Standing Committee on Examinations. While noting that there might be thirty students per year, and that sixteen potential teachers in Anthropology and Geography together with nine in other faculties who had 'promised their active assistance', it concluded that if a new School were established there would eventually be a need for accommodation and further teaching appointments (OUA, UR 6/ANT/4: Note, 2 May 1948). Moreover, the teaching for the proposed School would have to be done 'outside the Colleges'; that is to say, most – if not all? – of the proposed teachers were employed in museums, science departments, etc., and did not have undergraduate College Fellowships (OUA, UR 6/ANT/4: Report of meeting,15 October 1948). The General Board then consulted a range of faculty boards and other interested parties.

There were two real sources of difficulty in getting the proposal through. The first emerged from the Social Studies Faculty Board, which asked whether Anthropology might not more appropriately be examined as part of a wider School instead of a single School in itself, while the 'special subjects' should be at postgraduate level (OUA, UR 6/ANT/4: Letter from Secretary of Faculties to E-P, 11 December

1948). The second difficulty emerged from the Lit. Hum. committee. Their criticisms were based on a distinction between 'an education', which is the proper concern of an Honour School, and a 'technical training', which is the business of a Diploma course. They did not doubt the suitability of anthropology to have a place in an Honour School, but 'we are not convinced that a satisfactory education can be obtained from a School so predominantly confined as that envisaged by the present plan to the study of man in a primitive or uncivilised state' (OUA, UR 6/ANT/4: Communication from General Board to Faculty Boards, 28 October 1949: 5). In the end the Standing Committee on Examinations decided not to approve the proposed degree. They concluded that it seemed neither to provide a strict scientific training nor a humanistic education; that the material used would hardly ever be first hand, and could not be tested by undergraduates 'who will be driven to rely on opinions expressed in lectures'; that there was force in the points made by representatives of Lit. Hum.; and that the studies proposed were essentially postgraduate in nature (ibid.: 2–3). This was not the last time that difficulties for anthropology would emerge from rival interests, especially perhaps in Social Studies, as will be seen below.

Had the proposal for the undergraduate degree been accepted, the course of anthropology at Oxford would have been rather different. Staff would have had to concentrate a good deal more of their energy on the systematic teaching and administration of an agreed syllabus. This would have left much less time for encouraging innovation, promoting publications, and making the most of the interdisciplinary mix represented by staff and students – activities which, in the event, they were free to concentrate on.

The 1950s: a campaign for social anthropology

Social anthropology in Britain inherited, it goes without saying, a deep connection with the imperial world; this was particularly the case in Oxford. There was, for example, a direct and close collaboration with the Committee for Colonial Studies in the University. Institute staff devoted a good part of their time to lecturing Officers and Probationers of the Colonial Service (later known as Colonial Cadets, and then as Overseas Services Cadets). As Figure 9 shows, this commitment continued right up to 1962. At least one lecturer (John Beattie) and several students had formerly served as colonial district officers. E-P himself lectured to the Cadets on the relation of anthropology to colonial issues, insisting, however, as he did so on the autonomy of academic research – which was not merely a matter of providing useful information to administration (abstracts of the lectures survive; see E-P 1937b, 1938). However, in his more general teaching, writing, and publications, he was addressing a wider horizon. He was explicitly constructing a new agenda for social anthropology, speaking to older and broader questions about human nature and human history, shared with neighbouring disciplines in ways which would, in the event, help the subject survive the 'end of empire'.

In the – usually positive – assessments of the first decade or so of the E-P years it has often been forgotten how vulnerable social anthropology was as a University discipline at this time. The responsibilities of the professor were to lecture and supervise research in the discipline, and to have charge of the library. There was no established 'Department' in the modern sense, and as E-P himself explained: 'The "Tylor Library" forms the nucleus of the Institute of Social Anthropology. This fine library, the kernel of which is Tylor's personal collection, presented in 1911 ... not only provides full teaching facilities but also ... full research facilities for students' (1951b). However, there were very few professional teachers in any of the fields of anthropology, and their various efforts to establish the discipline, jointly or separately, faced firstly the inbuilt difficulty of change within Oxford and secondly the inevitable emergence of rivalries and intrigues among themselves. E-P's struggle to establish social anthropology in the 1950s faced competing interests, on the one hand, from the museum side and from biological anthropology, but also on the other hand from the ambitions of Sociology. By contrast, the Institute's best allies emerged from subjects that were not rivals, and which might have inherited some sympathetic attitudes from the Marett era: that is, in particular, philosophy and religious studies, where E-P's own friendly contacts with the religious community of Blackfriars were certainly fruitful. In this respect, it was a great advantage for social anthropology to be channelled solely into postgraduate training and research with a leaning towards the humanities. Outside Oxford University, the Institute of Social Anthropology enjoyed considerable success both in the local 'town' constituency (where there was a long-standing interest in anthropology, as shown in Parkin's chapter), and more importantly in the wider national and international setting. Both kinds of success helped it survive and eventually root itself firmly within the University.

The opposition encountered in 1949 to the idea of an Honour School helps us see how and why there developed a kind of crusading spirit in the Institute, to promote the cause of social anthropology despite the local sceptics. I first became aware of this as a first-year undergraduate in Geography, through the year 1959–60. We had lectures at the Pitt Rivers Museum, by Audrey Butt (later Colson) and

Figure 9. Institute Students 1946–73 (Source: Institute of Social Anthropology Annual Reports)

ACADEMIC YEAR	Colonial Officers	Colonial Cadets*	Regular students registered	Number sitting Diploma/Certificate‡	Number receiving B. Sc.	Number receiving B. Litt.	Number receiving D. Phil.
1946–7	4	70	20	?	1		2
1947–8	4		32	9	2		1
1948–9	4		54	8	2	1	1
1949–50	3		60	8	3	1	2
1950–1	3	30	50	12	3	11	3
1951–2	2	25	43	10	1	3	6
1952–3	2	22	37	6	3	6	8
1953–4	1	17	48	15		5	
1954–5	2	26	43	7		1	3
1955–6		20	43	2	2	5	3
1956–7		36	39	9		5	6
1957–8		45	41	8		2	4
1958–9		25	43	11	1	8	3
1959–60		28	42	13	1	5	1
1960–1		11	44	9		1	4
1961–2		15	58	17		3	3
1962–3			63	25		7	5
1963–4			78	29		5	4
1964–5			85	35‡		11	3
1965–6			112	34		11	4
1966–7			113	33		11	5
1967–8			118	25		17	9
1968–9			107	20		14	6
1969–70			118	24		17	2
1970–1			114	23		9	9
1971–2			122	30		9	6
1972–3			120	23		9	7

*From 1959, they are referred to as 'overseas cadets'.

‡Figures from 1946–7 refer to the general Diploma in Anthropology; but from 1965 to the Dip. in Social Anthropology only.

Ken Burridge, on something like 'Peoples of the World' with demonstrations of their material culture, and I found these riveting – especially my first introduction to the Nuer. This prompted me to request E-P's *Social Anthropology* as a school prize I was due. I started asking friends about the Institute; A.R. in my College told me 'they're all brilliant [in the sense of *madly* brilliant]; and they're mostly Catholics; the only sane one is John Beattie'; and because I was involved later in various African-related activities, I met Godfrey and Peter Lienhardt in the bar-billiards room at the Commonwealth (formerly Colonial) Services Institute. After sizing me up (I had just been in Tanganyika) Godfrey fixed me with an intent eye, pronounced on my character in a way that would have been difficult to accept from anyone else, and announced that I should come to the Institute.

Part of the promotion of social anthropology consisted of providing a history for the subject itself, tracing sources for its key ideas or what E-P once called social anthropology's 'theoretical capital'. Filial allegiance was accorded the Durkheimian school, and a whole string of newly discovered ancestors in the tradition of the French and Scottish Enlightenment was lined up as a genealogy of the discipline. An explicit break was made about this time with R-B's views; E-P's engagement with political processes and regional cultural connections had always been a feature of his work, and his wartime research on the Sanusi of Cyrenaica had taken on a truly historical character. It is not clear why his inaugural lecture of 1948 placed social anthropology squarely within the social science tradition we would now associate more with R-B, though they were jointly concerned (with Fortes) to give a shape to the subject during the 1940s. There is a view that the text of this lecture owed something to an earlier manifesto-style draft by R-B himself, who never gave an inaugural. Another received report suggests that the delivered version differed from the later published text (both these oral reports deserve further research). But E-P's Marett Lecture of 1950, delivered safely after R-B's departure, famously took a different turn and proclaimed that the subject was properly one of the humanities. E-P's personal vision, now licensed to re-invent social anthropology in the image of the 'humanities', was after all more in line with his original training as a historian, and he was now free to pursue it. He was very pleased to be invited to deliver a set of radio lectures in 1950 to a wider audience, later published (1951a) and translated into Japanese, Spanish, Arabic, Hindi, French, and Italian. A local manifesto was published in the *Oxford Magazine* (1951b), E-P here setting out a confident and optimistic agenda for the subject. A slim volume of essays was edited from further broadcast talks, several by Institute colleagues, as *The Institutions of Primitive Society* in 1954. An update on the progress of anthropology in Oxford appeared in *Man* (1959).

His field research over, E-P had turned to critical and reflective studies (cf. Gellner's introduction (1981) to the posthumous collection of his pieces edited by André Singer). In addition to writing up further field notes (mainly on the Azande) and reworking some of his existing publications into the volume *Nuer Religion* (1956), he produced a steady series of commentaries on his chosen predecessors and on topics such as the nature of religion. In doing this, he went far beyond the writings of paid-up members of the anthropological profession. His aim was parallel to the project of the philosopher and archaeologist R.G. Collingwood, who had provided critiques of 'the idea of nature' and of 'the idea of history', not to mention concepts of art and philosophy, by engaging with the way that thought patterns of the past have been periodically transformed (e.g. Collingwood 1939, 1945, 1946; cf. 2005).

Who or what may have helped E-P to select his newly chosen forebears in the series of lectures and papers he worked on during his time in the Oxford Chair? In my view, his thinking and reading on this matter was considerably influenced by the writings and conversation of Sir Isaiah Berlin, who had been a Fellow of All Souls from 1932–8, and then returned in 1950, not long after E-P's joining the College in 1946. There is, as far as I know, no reference to Vico, Montesquieu, Comte and so on in E-P's writings before the War, but these are standard points of reference in Berlin's lectures and essays, as they became in E-P's own revision of social anthropology's history.[4] E-P's friendship with Berlin was a steadfast one, and I myself remember Berlin being invited to give a talk on Vico to the Institute in, I think, 1969 or 1970. When Sir Isaiah became the founding President of Wolfson College, he was very supportive of anthropology.

The graduate students who came to Oxford were of varied and often mature experience, from the early days when a good number had wartime employment of one kind or another, on to the later years when not a few were doctors, priests, or professionals in fields such as translation or business. They came with existing qualifications in a range of subjects, mainly the humanities, and thus brought a variety of perspectives into their study of anthropological theory and ethnographic literature. The result was an enrichment of debates at the Institute, and of variety in the individual research contributions made. It is well known that Godfrey and Peter Lienhardt, and David Pocock, studied English with Leavis in Cambridge; it is less well known that among the staff John Beattie's first degree was in philosophy; or that Paul Bohannan had read German literature, Ken Burridge jurisprudence, Mary Douglas PPE (politics, philosophy and economics), Meyer Fortes psychology with English, Rodney Needham Chinese and Malay. Among students of the early 1950s, Freddy Bailey had read PPE, Paul Stirling Lit. Hum., John Middleton English, Ioan Lewis chemistry, John Barnes maths with anthropology, Clyde Mitchell sociology and psychology, Robert Paine history, and Peter Lloyd geography. In the later 1950s into the1960s, the pattern continued with a similar mix but expanded numbers. Still only a few came with existing degrees partly or wholly in anthropology. This diversity helps us understand why so many of the professional publications from the Institute were addressed to a readership way outside the confines of the anthropology classroom or teaching syllabus.

The literary activities of the Institute took a variety of forms. Translations of E-P's own works appeared in a range of languages, often sponsored or undertaken by students. A series of English translations from the works of the Durkheimian school was promoted by E-P shortly after his appointment. He approached Ian Cunnison (not long after his arrival as an Advanced Student in 1947 after studying French along with his archaeology and anthropology in Cambridge) about the work, and contacted a potential publisher about this project as early as April 1949.[5] The series was launched in 1954 with Cunnison's translation of Marcel Mauss's essay on *The Gift*.[6]

In 1956, another Oxford initiative saw the launch of the journal *Contributions to Indian Sociology*, edited by David Pocock of the Institute and Louis Dumont, whom he had succeeded, and who was now at L'École Pratique des Hautes Études in Paris. It is worth noting that the good relations cultivated by the *Times Literary Supplement* (from at least 1955) with anthropology owed something to the literary-philosophical style of anthropology that E-P himself helped foster; the anonymous reviews carried by the *TLS* included anthropology books of various provenance, though not infrequently they were reviewed by Oxford authors. Several pieces by Institute staff also appeared from time to time in *The Listener.*

Consolidation: the 1960s

The public pronouncements on social anthropology's 'progress' continued during the 1960s. Rodney Needham noted in *Man* (1964) how the old Diploma in Anthropology at Oxford was now split into four specialist Diplomas, of which by far the most in demand was that in social anthropology. The habit of providing reports to the general public continued right through E-P's tenure of the Chair and beyond; in 1973, when the ASA's Decennial Conference was held in Oxford, the *TLS* carried a final piece by him on 'Fifty Years of British Anthropology'.

Intense efforts were continued right through E-P's tenure by himself and his colleagues to define the scope of what they understood as a subject quite transcending what were seen as older kinds of anthropology. There can scarcely have been a department in which so many close colleagues published introductions to their discipline, or polemical or promotional arguments aimed at specifying its focus. E-P's own 1951 book and programmatic writings from the 1950s were followed by a spate of others in the 1960s. The first of the introductory books was Pocock's *Social Anthropology* of 1961, followed by Paul Bohannan's *Social Anthropology* of 1963 (though I do remember E-P being a bit dismissive of the latter); and in 1964 both Lienhardt's *Social Anthropology* (the final fruit of an original invitation by the Home University Library to R-B) and Beattie's *Other Cultures: Aims, Methods and Achievements of Social Anthropology* appeared (the main title having been proposed by the publisher's US branch). Needham's *Structure and Sentiment*, a passionate argument for the rule-based nature of social relations, and thus a defence of social anthropology against psychological explanations, appeared in 1962. Mary Douglas (who of course had long since moved from the Institute to UCL) published *Purity and Danger* in 1966; this certainly drew very wide attention to a style of anthropology many would associate with her Oxford background, and in its own way constituted an important extension of anthropology's conversations with philosophy and theology.

The crusading ethos continued through the whole of the E-P period and indeed beyond – it resonates with some of us still (see Dresch and James 2000: 4–6). Let me just mention here Edwin Ardener's Malinowski lecture of 1971 which was entitled 'The New Anthropology and its critics', and constituted a broadside on scientism, as did his elegant short piece '"Behaviour": A social anthropological criticism', published only a couple of years later, aimed mainly at students – and even fellow teachers – in the new Human Sciences degree (Ardener 1971, 1973). It was, significantly too, published in the *Journal of the Anthropological Society of Oxford*, over whose emergence Ardener had presided in 1970. Polemical and critical work by Oxford-trained anthropologists was also spreading in other directions: let me just mention the conference envisaged in the late 1960s and organised in 1972 by Talal Asad, then at Hull University after a period of teaching in the Sudan, which demonstrated the formative effects of the colonial period upon anthropological thought and practice. He had tried to get the ASA to adopt this topic for one of their annual conferences, but it was turned down by Raymond Firth, reportedly, on the grounds that 'we went through all that in the 1930s'. The resulting volume (Asad 1973) has attracted enormous attention.

Social anthropology during the post-colonial years did start to catch the eye of scholars in neighbouring fields. A notable example of the interchanges which took place were the 1960s debates on 'rationality' with philosophers (Wilson 1970), continuing later with joint meetings on topics like 'sacrifice' with the theologians (Bourdillon and Fortes 1980). Some had forecast that the demise of colonialism would put an end to the discipline; but the opposite proved to be the case, as there was a great expansion of student interest from the early 1960s (see Figure 9 again). The growing ranks of graduate students more than replaced the trainee administrators. When I myself became a student at the Institute (in 1962), I think it would not be wrong to say that the excitement of rising nationalism and colonial liberation in Africa and elsewhere provoked a new and potentially radical interest among young people, in the UK at least, about what anthropology had to offer as insight into the contemporary peoples and cultures of such regions, and thus the events going on in them, since conventional books on history or politics had very little to say. Reciprocally, students from Africa and other 'remote' parts of the world often found an affinity with the anthropologists because so few other academic disciplines actually focused on grass roots studies in these countries.

The research agenda which E-P was promoting included from the start the study of areas beyond the contemporary colonial territories. He had been able to establish a post in Indian Sociology as early as 1948, and to push for the development of Mediterranean anthropology, which John Campbell was able later to consolidate over many years from his position at St Antony's College. Between 1960–3, the years which finally saw the end of the Overseas Cadets and showed the new trend, completed B.Litt. theses included Finnegan on early Irish kingship, Asad on changing family structures in the Punjab, and Kesby on British missionaries in the Pacific; and D.Phils. included Paine on the coastal Lapps of Norway, Miller on the religious kibbutz, and Maraspini and Lisón-Tolosana on rural Italy and Spain respectively (Institute of Social Anthropology, Annual Reports). There were also of course the more standard anthropological-looking field studies in Africa, South America and S. E. Asia. Studies of this kind seemed increasingly, however, to reach into questions of history, cultural change, oral literature, and religion; and indeed politics, as radical questionings of the older forms of authority and the dominance of the West began to spread. There was no great taking to the streets in Oxford, as in Paris and London, or campus occupations as in the United States, but the Proctors wisely turned a blind eye as gowns gradually disappeared from lecture rooms. 'Anthropology at home' was given a good start in Oxford in 1968, with Barbara Harrell-Bond's study of the Blackbird Leys housing estate on the outskirts of the city's declining industrial zone of steel and motor manufacture.

On the literary front, there was a burst of activity in the 1960s. Relations between the Institute and Oxford University Press were very good during E-P's time. He had proposed a series of Oxford Monographs in Social Anthropology (securing a grant of £500 from the University) in 1956. The series, overseen by a committee including E-P and John Peristiany of the Institute, Tom Penniman of the Pitt Rivers Museum, and Isaac Schapera of the LSE, was launched in 1963 with Jean Buxton's first book on the Mandari, and this was followed up with fifteen other titles by 1979. Several were the distinguished results of doctoral fieldwork (for example, Cutileiro 1971, Gilsenan 1973, Barnes 1974). However, it was the normal practice for graduate students to prepare theses mainly based on library work for the degree of B.Litt., before going to the field for their doctoral research, and the standard of these was often very high (a valuable output not paralleled, to my knowledge, in other departments). Several B.Litts. became books in the Monograph series, for example Eva Gillies on Yoruba towns and cities (Krapf-Askari 1969), Hilary Callan on the relations between the study of animal and human behaviour respectively (1970), and Hilary Henson on the historical relationship between British social anthropology and language (1974), while other B.Litts. of this period also found willing publishers (for example, Just 1989).

As part of the engagement with the humanities, E-P with Godfrey Lienhardt and the linguist Wilfred Whiteley launched a major and brand new series, the Oxford Library of African Literature, with four volumes in 1964. By 1979 when it came to an end, some 27 volumes of largely 'unknown' stories, oral and literary genres, songs, and myths had been published, an extremely significant contribution to the growing field of the study of African history, languages, and civilisation. Many involved in producing the series were trained originally in Oxford – for example, Ioan Lewis (1964 with B.W. Andrzejewski), Ruth Finnegan (1967, 1970), Peter Lienhardt (1968, edition of Hasani bin Ismail), Jack Goody (1972), Roy Willis (1978), Ahmed al-Shahi (1978 with F.C.T. Moore); but the majority of the volumes were edited by scholars from elsewhere – including several from Africa itself, such as John Mbiti (1966), S.A. Babalola (1966), Daniel Kunene (1971), and Francis Deng (1973).

There was rarely a conscious sense of a party line within E-P's Institute, especially as internal rivalry developed, sometimes reaching the public domain (see for example the articles by Needham and by Beattie in the journal *Africa* during the 1960s). But beneath the competing efforts to define, explain, and defend the principles of social anthropology and somehow capture its soul, there was much in common. Looking back we might sum up its approach as a comparative but empathetic study of the forms and experiences of human sociality, in so far as we can have access, as scholars and as persons, to their historical reality. The currents of new systematising theory which spread in the 1960s, in particular Lévi-Strauss's structuralism, French neo-Marxism, and the new utilitarianism ('rational choice') could be met and engaged with; as just a little later, social anthropologists began their efforts to engage with the challenge of the new biological sciences through the Human Sciences degree. But the most comfortable engagement, as I see it, was yet to come, in the context of the Archaeology and Anthropology undergraduate degree, which began only in 1992.

Institution building

The Institute and its Chair survived the E-P years partly by design and hard work, but partly also by hazard, and by odd favourable winds. It was surely an asset to the rest of the University because of its outstanding reputation and ability to attract students, but administrators and committee chairs probably felt it better not to provoke the sometimes fiery loyalty of its denizens. E-P certainly worked hard to build up the Institute's finances, its posts, and its accommodation, through the move from 1 Jowett Walk (shared with the geographers) to Museum House in South Parks Road, then to 11 Keble Road, and then in 1965 to 51–3 Banbury Road. The story of the accumulation of posts has been summarised by Rivière in the introduction to this book; here let me just emphasise the support of the Colonial Studies Committee and of Oriental Studies, especially for the posts in Indian and in African Sociology, first held by Srinivas and by Lienhardt respectively. These were finally transferred to the establishment of the Anthropology and Geography Faculty Board in 1966, which, as Rivière has pointed out, in retrospect was an addition which helped to secure social anthropology during the retrenchments of the following decades – and as I note below, their very titles were judiciously redefined a few years later when the social anthropologists felt the need to close ranks against poachers.

Many of the immediate post-War students were able to obtain funding from the Colonial Office or win Treasury Studentships. Others in the 1950s were supported, especially for field research, by a great variety of bodies – including the Rhodes-Livingstone Foundation, the Sudan Government, the Wenner-Gren Foundation, the SSRC (USA), the Horniman and Leverhulme Funds, Goldsmiths, All Souls College, the International African Institute, the Egyptian and Belgian Governments, and the Australian Army Fund (E-P 1959: 121–2). Philip Bagby, a former student, left funds for the comparative study of the development of urban, literate cultures in accordance with

Figure 10. The Institute staff in the library at 11 Keble Road in 1953. Standing, left to right, Godfrey Lienhardt, Paul Bohannan, David Pocock. Seated, left to right, Phyllis Puckle (librarian/secretary), E-P, John Peristiany. Copyright ISCA, University of Oxford.

anthropological principles and methods; the studentship in his name was first advertised in 1963, and has regularly supported a student since. As E-P's retirement approached, colleagues proposed to set up a fund in his name, and he asked that it commemorate his late wife; the Wenner-Gren foundation matched the monies raised locally, to provide the Ioma Evans-Pritchard Studentship at St. Anne's College (later a Junior Research Fellowship). In the 1960s, the British SSRC was generous to social anthropology; between 6 and 9 studentships were provided to Oxford each year from 1966–9 and this rose to 13 in 1970.

The archives contain a surprising amount of paperwork over specific efforts by E-P to improve the infrastructure and equipment of the Institute, such as to secure a £120 grant for a tape-recorder (1954) and then £25 for a photocopier (1955), not to mention regular efforts to persuade the various levels of the administration that further secretarial or academic posts were needed. He was treated with some circumspection, I think, partly because of a kind of personal doggedness – one complaint against his behaviour (by John Layard, Oxford resident and anthropologist of Melanesia, trained in psychoanalysis and of considerable emotional temperament himself, who wished to make use of the Institute) reached the Visitatorial Board in 1958. While E-P's decision to exclude Layard by rather firm means was upheld, this case cannot have made E-P or his colleagues seem any easier to deal with. It also revealed some of the uncertainties over the extent of the formal, statutory powers of the Professorship of Social Anthropology, and the activities over which he presided (as mentioned, these were formally limited to lecturing and supervising research in social anthropology, and having care of the Tylor Library). E-P's demands tended to be treated with care by the University administrators, some of whom were aware of one or two anomalies about the standing of the Chair and the Institute. In 1960 E-P claimed that he was the only Professor with administrative responsibility who did not draw an administrative allowance; an internal comment in the files from an administrator notes 'I realise that the existence of the Institute is *de facto* and not *de jure*' (OUA, UR 6/ANT/1 file 2).

From the statutory point of view, at that time 'Departments' were defined essentially on the 'science' model, and specific professors had charge of specific departments, thus entitling them to an allowance. In 1964 a proposal to use 'Department' was turned down by the General Board. They were not convinced this was necessary, that it would raise a number of difficult questions and require legislation, and that it would be better for the matter to be left in *statu quo* (OUA, UR 6/ANT/1 file 2 [AG(64)42]). The University saw the Institute not as a department but only as a 'unit of academic administration'. If social anthropology could get along without being a statutory Department, so be it.[7] The uncertainties felt in the Institute over the University's attitude were matched, no doubt, on the other side by a perception of E-P and some of his colleagues as unpredictable, and there is no doubt of the spreading reputation for eccentricity.

The Institute's achievement is nevertheless beyond questions of personality or quirkiness in E-P or for that matter his colleagues. He was a prolific scholar throughout this life; a rough count based on Beidelman's bibliography (1974) gives for the 1920s, sixteen published items; and for subsequent decades, 58; 82; 70; 84; and 74 (just up to 1974). While the position of social anthropology was remarkably fragile in Oxford itself, the Institute under E-P was well known as a nursery for the training of professional anthropologists who spread out to take posts and found departments in other UK Universities and across the globe. Students, staff, and academic visitors were attracted from overseas to a remarkable extent, many then returning to develop their own networks, and many keeping in touch with Oxford through correspondence and reciprocal invitations.[8] As E-P himself noted in 1959, 'Being a post-graduate school, we are largely dependent on recruitment from outside Oxford and partly from outside the British Isles' (1959: 121) (a dependence, in both respects, which continues to the present, and in respect of overseas recruitment has greatly increased).

The threshold of an era: 1970 and the succession

At the end of the 1960s Oxford anthropology was flourishing. Taking the information in the very full loose-leaf edition of the ASA Annals for 1969, I find that of the total of 168 members of the Association, forty-two had acquired either the B.Litt., or the D.Phil., or both, in social anthropology at Oxford – that is, precisely a quarter. For those with doctorates in social anthropology at other Universities, the LSE came closest with thirty-seven. UCL had eight, SOAS one and there were seven described as simply 'London', so taken as a whole London would predominate. Others of note were Cambridge with sixteen, Manchester with fourteen, ANU with nine; but no other had more than four.

However, as was perhaps only to be expected, E-P's charisma had led to a dilemma for the succession. He had encouraged a very high level of disciplinary commitment among his students and colleagues, and it can easily be understood that there was no internal or external candidate obviously acceptable all round as the successor. Quite apart from this, however, there is evidence that the social anthropologists were justified in some of their fears about the future of the Chair. The

Social Studies Board, which as I have noted made some observations in 1948 about the need for anthropology to be placed in a wider setting, reported to the General Board for its meeting on 17 July 1968 that it considered the occasion of the vacancy in the Chair 'might be an opportune time for reconsidering the general position of Social Anthropology in the University, particularly in relation to Sociology; [and] that the electors should be prepared to consider applications from persons interested in modern industrial communities as well as primitive societies and that the post should be advertised in terms that would make this quite clear' (OUA, UR 6/ANT/2A, file 2). However, the anthropologists proposed that 'The study of Social Anthropology should continue to be advanced along the lines developed by the present Professor', which led to the Board stating that they were willing to discuss the advancement of sociology as such, but not in relation to the Professorship of Social Anthropology. The A&G Board dragged its feet about who their representatives should be for the joint meeting to the point where the Secretary of Faculties reported that the only practical course would be for the Planning and Development Committee to see representatives of the two Boards separately, '(since otherwise there might be an explosion dangerous to life)'. The Committee later reported that the Social Studies Faculty Board 'certainly did not have in mind a take-over', but there was a close connection between social anthropology in some of its aspects and sociology; 'the point really was that the line of work in which Professor Evans-Pritchard has made his uniquely distinguished contribution had perhaps been taken as far as it could profitably go, and the Social Studies Board did not want to see him replaced merely by a "disciple" who would simply follow in his footsteps ...'.

In the event the General Board authorised the filling of the Chair without making any appeal to 'persons interested in modern industrial communities as well as primitive societies'. The battleground chosen by the sociologists was based, surely not in all innocence (though that might have been a reasonable excuse in 1949), on an outdated image of what social anthropology had by this time become. The work of E-P and the Institute over the period since the War had surely modified, if not transcended, the old conception of the 'primitive' as its object. But there were other problems, ironically, in relation to what had evolved from Colonial Studies and was by this time Commonwealth History, whose representatives, including the distinguished figure of Margery Perham, also looked rather askance at social anthropology, so little concerned with 'progress' (I know this from an unsuccessful interview I had at Nuffield with her).

During the intensive period of the 1950s and 1960s, social anthropology had defended itself well against doubters and critics in the University, but its devotees under E-P had developed the virtue of loyalty almost to a fault. They were seen variously as creative and brilliant, sometimes jealous of each other, unpredictable and difficult to manage, more awkwardly defensive by 1970 in relation to the history of empire than they had been during its heyday. They did manage to hang on to their Chair but, arguably, with crucial support from external champions of the subject in other Universities rather than wholehearted endorsement from within Oxford. The succession went in 1970 to an unexpected 'outsider': Maurice Freedman, from the LSE, was a specialist in China and the overseas Chinese communities, an ethnographic field quite new for Oxford.

John Beattie left to take up a Chair of African Social and Cultural Anthropology in Leiden, being succeeded by Peter Rivière who was able to consolidate South American anthropology. Rodney Needham resigned in order to accept a Chair in the University of Virginia, though he later withdrew from this move (finally being elected to the Chair in 1976 after Freedman's premature death). Concurrent with preparations for the 1970 succession, plans for the new Honour School of Human Sciences were coming to fruition, spearheaded on the social anthropology side by Edwin Ardener, and by Geoffrey Harrison on the side of biological anthropology, with a number of colleagues from other fields. Institute staff found various ways to further protect and prolong the life of social anthropology in the University: for example, by retitling the posts held by Ravi Jain and Godfrey Lienhardt from 'Indian Sociology' to 'The Social Anthropology of South Asia' and from 'African Sociology' to 'Social Anthropology' *tout court*. Within a short time of Freedman's appointment, further accommodation had been acquired at 61 Banbury Road, and a brand new lecturership in social anthropology secured (to which I was appointed in 1972). E-P was knighted, but only after his retirement; his vision had in some ways run its course, but in other ways it has left a permanent and valuable legacy. This is why so many evoke even now the myth, the special aura, of the E-P years. He would have been delighted to know that in 2005 we were able to secure a second chair in the discipline he championed with such infectious passion.

Notes

1. In preparing this chapter I have been given guidance on earlier drafts and helpful suggestions by Eva Gillies, Richard Fardon, Bob Parkin, Peter Rivière, and Malcolm Ruel.
2. Jack Goody spent part of his student days at the Institute, taking the B.Litt. in 1952; his study of the rise of British social anthropology devotes a chapter to 'the

Oxford group' (Goody 1995: 77–86). John Barnes has also written on E-P's anthropology from the point of view of a one-time student (Barnes 1987). John Burton, trained in the USA, carried out fieldwork in the southern Sudan in the 1970s and was a frequent visitor to Oxford; he has provided several assessments of anthropology at Oxford (for example, Burton 1992). Godfrey Lienhardt arrived in Oxford from Cambridge (initially registering for the D.Phil.) in 1948 and became a close colleague of E-P's.

3. Mary Douglas, in her maiden name Tew, joined the Institute in 1946 as a B.Sc. student, gaining this degree in 1948 and her D.Phil. in 1952. She served as a lecturer between 1950–1 before moving to UCL. For her study of E-P's ideas and their roots in what we might call the cognitive disciplines, see Douglas (1980). Richard Fardon's contribution to the history of Oxford anthropology includes not only his full-length intellectual biography of Mary Douglas herself (1999), a book which includes a quite complex portrait of the 1940s and 1950s (especially Chapter 2 on 'Oxford Years: 1940s'), but also his three co-edited volumes devoted to the work of Franz Steiner, who became a key figure in that setting even before his premature death in 1952 (Steiner 1999a, 1999b; Adler, Fardon and Tully 2003).

4. On a hunch I compared the indexes of the volume edited by Singer (E-P 1981) and Berlin's *The Hedgehog and the Fox* (1953). There are no fewer than twenty-one authors occurring in both lists, and this number includes some quite new names for anthropologists to ponder adding to their genealogy – such as Condorcet, Diderot, de Maistre, Chateaubriand, Sorel, Taine, Voltaire.

5. Letter from E-P to The Free Press, Glencoe, Illinois, of 20 April 1949, mentioning that Cunnison had partly finished a translation of some well-known essays by Mauss and by Hubert and Mauss – 'The Gift', 'Magic', and 'Sacrifice', asking if they might be interested in publishing them. The Free Press had already brought out some translated work of Durkheim, and replied (9 May 1949) that they had already been considering an English translation of some of the more interesting papers by Mauss, and expressed interest in seeing the manuscript (OUA, ISA F/I, 25). In the event, the main publisher for the series became Cohen and West.

6. A clutch of other significant texts followed, and the tradition of these translations is continued at Oxford to the present day, with the support of the British Centre for Durkheimian Studies-the most recent are versions of Hubert's *Essay on Time* (1999), and two works of Mauss, *On Prayer* (2003) and *Techniques, Technology and Civilisation* (2005).

7. This matter of course did come up again, and did have financial and other implications; it seemed to be resolved in 1977–8, after the title of 'department' came to be used across the University in a more varied way than before. At that time, the Institute adopted a constitution based upon a departmental committee with a rotating chair, a pattern already obtaining in the Mathematical Institute. By the late 1980s, however, the Chairman of the General Board himself proposed the adoption of the title 'Department', in the context of the planning of the new School of Anthropology; Peter Rivière, then on the General Board, prevented the change on the grounds that you do not change a successful brand name (Rivière, per.com.).

8. This is clear from the files of correspondence from the 1950s and 1960s, often containing significant sets of letters from students in the field and in posts elsewhere (OUA, ISA/F, Correspondence files).

Chapter 6

OXFORD AND BIOLOGICAL ANTHROPOLOGY

Geoffrey Harrison

Introduction

Physical Anthropology, later to be renamed Biological Anthropology, was born out of Darwinism and evolutionary theory. The fact that there was a continuous descent connection between human beings and the rest of the living world added an important biological dimension to the study of human existence. Initially, the subject was almost exclusively evolutionary in content and concerned with the course of human descent as evidenced in the fossil record, the comparative structure of humans with their closest living relatives: monkeys and apes, and the great biological diversity of recent human populations, which apparently could be classified into races. Such studies required detailed knowledge of human and comparative anatomy and particularly of bones and it is therefore not surprising that physical anthropology was mainly studied in anatomy departments of university medical schools. This was very much the case at Oxford where for well over a half of its history physical anthropology was a laboratory of the Department of Human Anatomy, where two professors of Anatomy, Arthur Thomson and Sir Wilfrid Le Gros Clark, not only promoted the subject but also taught and researched in it.

After the Second World War, however, the subject changed quite markedly and became much broader. While the focus has remained evolutionary, attention was increasingly given to all aspects of human and primate variety, its nature, origins, development, causes and effects at the cellular, tissue, organ and whole body levels including behaviour and with as much attention given to environmental determinants as to genetic ones. Thus the subject now includes human population genetics, human developmental biology, physiological anthropology including nutrition, medical anthropology and human ecology sometimes grouped as human population biology but more commonly included with the evolutionary perspectives as biological anthropology.

Oxford was among the first institutions in the world formally to make the name change of physical anthropology to biological anthropology: a change that was recognised by the *ad hominem* promotion of Geoffrey Harrison to a Professorship in Biological Anthropology and the establishment in 1976 of a Department – later termed Institute – of Biological Anthropology, fully independent of Anatomy and housed in an attractive old Victorian house on the Banbury Road. On Harrison's retirement the Professorship became established and the first holder, Ryk Ward was appointed in 1994. (A list of all the academic teaching staff in the subject is appended at the end of this chapter.)

Teaching

Throughout its existence the Diploma in Anthropology mainly attracted graduates interested in social anthropology. However, there was always a compulsory component of physical anthropology in it as well as optional papers. Indeed a number of senior biological scientists were on the Committee for setting up the course including W.F.R. Weldon, Professor of Zoology and F. Gotch, Professor of Physiology, as well, of course, as Professor Thomson. The syllabus for physical anthropology was given as 'Elements of Physical Anthropology including the comparative study of the principal anatomical characters which (a) determine the zoological position of Man amongst the Anthropomorpha and (b) distinguish the chief races of Man from each other together with methods of measuring and recording such characters'. Thomson offered lectures on the 'Elements of Physical Anthropology', Weldon on 'Elementary craniometry' and Gotch on 'Structure of Special Sense-Organs and Skin'. Apart from an ever greater focus on fossil remains as more were discovered, the course does not seem to have changed very much in the 1920s and 1930s.

The situation became very different after the Second World War. Generally this can be recognised as being due to the rise of genetics and greater understanding of the evolutionary processes, but in Oxford it can be attributed to Joseph Weiner. Weiner had experience in palaeoanthropology from Raymond Dart in South Africa, but he was first and foremost a medically trained human physiologist. He appreciated the significance of advances in genetics and was particularly interested in the nature of physiological adaptability. With Derek Roberts he substantially changed the composition of the Physical Anthropology syllabus of the Diploma. When Harrison succeeded Weiner, time was clearly right for further change and he, with John Owen, co-operated with other anthropologists and archaeologists in setting up in 1964 four separate Diplomas, for one of which, the Diploma in Human Biology, they were solely responsible (but see Introduction to this book, p. 9). The syllabus for the nine-month course consisted of human genetics, human development and behaviour, human ecology and human and primate evolution, each represented by a paper in the final examination along with a practical examination. It will be noted that only one paper focuses specifically on evolution which contrasts strikingly with earlier versions. It is sometimes asked why it was called 'human biology' especially as this has caused some confusion with anatomist, physiologists, *et al.* It has to be admitted that 'marketing' played a significant role, especially for a 'diploma' since this title carries little weight outside the UK. It was thought that human biology would 'open more job opportunity doors' and the subsequent experience of students confirms this. Human population biology would, however, have been a better name.

In 1979, a year-long M.Sc. in Human Biology was also established. The syllabus for the theory papers was almost identical to that for the Diploma but the Master's also required the submission of a dissertation of supervised original research. Following Ward's appointment some changes to the syllabus were made emphasising the increasing importance of genetics, and particularly molecular genetics, at the expense of human ecology.

After its establishment the Master's course became the preferred choice of almost all candidates. These tended to average about six each year, which was ideal, but occasionally as many as twelve were admitted and this taxed laboratory facilities to the limit. The M.Sc. became a common step but not formally a pre-requisite for admission as a D.Phil. research student. Today a number of steps are required before full admission as a research D.Phil. student and a Master's qualification either by research or course work counts towards some of these. Until very recently physical/biological anthropology at Oxford was considered to be best taught at the graduate level and candidates for the Diploma/Masters in Human Biology typically had first degrees in either biological or medical sciences.

Notwithstanding this focus on graduate training an optional paper in Anthropology was introduced in the Honour School of Natural Science as early as 1885! Initially the syllabus was quite wide and foreshadowed the Diploma (see Rivière, this book) but increasingly focussed on the physical anthropology elements. For much of its history, no one appears to have elected to offer it in examination but it began to be 'found' by a few students in the 1970s. Latterly, and through the efforts of Vernon Reynolds it became quite popular, especially with undergraduates reading chemistry and biochemistry. It also attracted a few zoologists some of whom later turned to anthropology.

Lectures in biological anthropology have also been offered at various times to medical students but the subject was never formally examined. Nevertheless, the lectures were usually well attended which, considering the heavy load of compulsory work for medical students, was gratifying. Some students even thought that the lectures would be of considerable value to them in medicine!

The main introduction of anthropology to undergraduate studies came, however, with the first intake for the Honour School of Human Sciences in 1970. The primary aim of this degree is to educate students with either scientific or humanities backgrounds on the inter-relationships between relevant biological sciences and social sciences: to bridge, as John Pringle the instigator of the degree saw it, the 'two cultures' divide. Clearly, a large anthropological input was needed and biological anthropology played a seminal role in the setting up of the degree and in its subsequent administration and teaching. Of the five compulsory papers in the Honour School, two and one-half have been mainly taught by biological anthropologists, Harrison, Anthony Boyce and Reynolds, and in recent times by Ward, Nicholas Mundy, Stanley Ulijaszek and Ros Harding.

Initially, there was hostility to the degree in some quarters of the University and some colleges, it being seen as too broad. But the breadth is something of an illusion and there is plenty of depth as has now been largely recognised. At least it only deals with one organism unlike the millions studied in zoology! From a modest start there is currently an annual intake of forty to fifty, whose school qualifications are at least as good as any other group entering Oxford. Gratifyingly graduates have found jobs in every kind of occupation from the commercial to the academic and are not seen, like most scientists, as being highly specialised.

A more specifically anthropological undergraduate degree was also set up with archaeology in 1992 – the Honour School of Archaeology and Anthropology. The anthropological component of this is heavily weighted towards social anthropology and ethnography, but there is an obligatory component of biological anthropology and a number of options in this field. Reynolds was much involved in the setting up of the degree and Ulijaszek now teaches the compulsory human evolution and ecology component.

All these courses, and many others elsewhere, benefited greatly from the publication of a text book, *Human Biology: An Introduction to Human Evolution, Variation and Growth* which among many things clearly defines the field of modern biological anthropology. The first edition of this book appeared in 1964 with two of the four authors closely associated with Oxford (Harrison, Weiner, James Tanner and Barnicot 1964). It subsequently ran to two further editions with some change of authorship but always with a strong Oxford connection and throughout published by Oxford University Press.

Research

No specialist palaeoanthropologist has ever been appointed in Oxford, but the University established a strong international reputation in the field. This was mainly due to Le Gros Clark, Professor of Human Anatomy. Although his principal researches were in neuroanatomy he had a career-length interest and commitment to anatomical studies of human evolution. He published two highly influential books; *Early Forerunners of Man* (1934) and *The Fossil Evidence for Human Evolution* (1955) essentially covering the whole field as then known, and he himself actively researched a number of issues. He was not one who went out to discover new fossil remains, which allowed him to be totally impartial in his evaluations, and instead he undertook wide analyses of major issues. He was not only a fine anatomist but also had a better understanding of broad evolutionary principles than any other anatomist of his time. His main accomplishment in this area was in establishing the status of the australopithecines. By a detailed quantitative analysis of functional units in the skulls of these organisms he was able to demonstrate unequivocally that despite their small brains and ape-like jaws they were terrestrial bipeds with clear hominid affinities. This confirmed the more impressionistic evaluations of their discoverers, Raymond Dart and Robert Broom in South Africa.

A very different kind of contribution to palaeoanthropology was the identification of the Piltdown fossils as fraudulent in the 1950s by Weiner. These remains, 'discovered' early in the twentieth century, became increasingly anomalous as more hominid fossils were found. They appeared to show that brain enlargement was the first feature to occur in hominid evolution. This was a major cause for the unwillingness of most British palaeontologists to accept the australopithecines as hominid. On learning that the provenance of the so-called second Piltdown remains was poorly recorded (which also showed ape jaw-like features with a modern human braincase, and thus could not be an accidental association which many people at the time had thought), Weiner concluded that the fossil must be a fake. In collaboration with Le Gros Clark and Kenneth Oakley of the British Museum (where the fossil was kept), Weiner showed on innumerable tests that the cranial pieces were those of an anatomically modern human and the jaw and teeth were those of a 'doctored' recent orang-utan (Weiner 1955).

A third less dramatic contribution to evolutionary studies was made by Oxford anthropologists in the early 1960s. Up to that time animal taxonomies, often had varying purposes, and often were merely impressionistic. 'A taxon was what a good taxonomist recognised as a taxon!' In particular, there was wide confusion between a natural (now called phenetic) classification and an evolutionary (or phyletic) classification. To try and help to overcome these problems systems of numerical taxonomy, based on the combined analysis of many characters, were developed. Harrison with Arthur Cain (of Oxford Zoology) contributed to this development and in particular showed how a phenetic classification could be converted to a phyletic one, and Boyce, an Oxford zoology graduate, was among the first to apply numerical methods to the classification of hominids.

Notwithstanding these significant contributions to palaeoanthropology, Oxford physical/biological anthropology has mainly

Figure 11. Joseph Weiner, Reader in Physical Anthropology 1945–63. Copyright School of Anthropology, University of Oxford.

focussed on modern and recent populations of primates and especially *Homo sapiens*. Initially, the concern was almost entirely with the affinities of groups of people with one another, with their classification into so-called races and with establishing historical connections between past and present groups. Since past groups are only represented biologically by skeletal material, a great deal of attention was given to bones and especially the skull which is a complex structure showing considerable variation within and between human groups. Comparisons were made by meticulous measurement and sophisticated statistical treatment. Indeed statistical science owes much to the skull measurers and if one declared in the 1920s and 1930s that one was a physical anthropologist, the invariable response was 'so you measure skulls then'. The fundamental weakness of the approach was that the skull was only seen as a complex piece of solid geometry and measurements were taken from points that were most clearly identifiable rather than of any functional significance. Nevertheless, the patterns of population groupings which were identified from skull and head measurements are quite similar to those derived from modern genetic analysis.

The first Reader in Physical Anthropology at Oxford, Leonard Dudley Buxton undertook some craniometrics and was, for example, involved in examining archaeological material from Crete and Mesopotamia. But he never became a slave to the approach like so many of his contemporaries. He was much more interested in general ethnology and recording the patterns of human variety around the world. He wrote for example a book, *The Eastern Road* (1924), covering his anthropological observations while travelling in the Far East, and spent time measuring the anthropometric features of Oxfordshire villagers. He had been a Demonstrator in Anatomy before his appointment as a Reader and collaborated with Thomson, in a number of researches. A particularly insightful piece of work they undertook was to examine the global distribution of variation in the nasal index. This showed a high correlation with the variability in the geographical distribution of atmospheric relative humidity and was perhaps the first occasion when anthropometrics were examined in a functional way.

Craniometrics became a pursuit of the past at Oxford when Weiner was appointed to the Readership after the Second World War. As a physiologist Weiner was primarily interested in function and how functions varied in different peoples in different environmental situations. His special interest was in the ways that humans adapted to high temperature and he researched many aspects of sweating mechanisms, which are so remarkably well developed in humans and must have played an important role in human evolution by allowing high levels of activity in the heat of the tropical day. He also refined the analysis of Buxton and Thomson on nasal structure by showing that geographical correlations were greater with better measurements of humidifying power than relative humidity. Clearly the nasal passages are adapted to ensure satisfactory humidification of inspired air. Weiner's interests were, however, much wider than human physiology. In addition to the Piltdown work already mentioned, he led a large expedition to South West Africa (now Namibia) to document the physical characteristics of the various ethnic groups found in that country, from blood groups to adiposity and skin colour.

During his time as Reader he headed a Medical Research Council (MRC) research group in climatic physiology, which on his resignation from Oxford he took to the London School of Hygiene and Tropical Medicine. He also became greatly involved in ergonomics. As just one of many examples, his team developed a 'bed' for the treatment of heatstroke victims on the Mecca Haj. He was the inspiration behind the foundation of the Society for the Study of Human Biology (SSHB) and the creation of its journal *Annals of Human Biology*. He also became the International Convenor of the Human Adaptability section of the International Biological Programme (IBP) – a major undertaking for which many countries were encouraged to develop their human population biology researches. Special emphasis was given to multidisciplinary approaches which often meant scientists from different countries collaborating and the standardisation of various measuring techniques.

Weiner's climatic interests encouraged his colleague, Roberts, to undertake a large scale analysis of the relationship between climatic variation and various body characteristics as recorded in the literature. Roberts was the first to show that the human species obeyed the ecological rules of Bergmann and Allen in that, on average, the surface area of the human body declined in relation to volume with increasing latitude and decreasing environmental temperature.

Roberts subsequently turned his attention to human population genetics. He undertook a pioneering study of the genetic structure of Dinka villages in the Sudan. This highlighted the importance of size and mobility on population structure. He also (with Robert Hiorns of Oxford Biomathematics) made some of the best estimates of the extent of the intermixture between whites and blacks in the USA. He later became Professor of Medical Genetics at Newcastle.

Harrison, with a background in biological and medical sciences at Cambridge and Oxford, had also been a student of Weiner and had similar interests: variation, adaptation and fitness in present day populations. His first researches, suggested by Weiner, were with experimental animals in which genetic factors could be controlled and all within strain variation was due to environmental conditions. He was particularly interested in features of growth and development and showed with mice in different temperature conditions and later, with John Clegg (of Liverpool), with rats at different atmospheric pressures that some of the environmental effects facilitated survival in the environments that produced them i.e. were adaptive, whilst others represented the health and well-being of the animals i.e. were fitness measures. Owen continued this animal work especially in unraveling the development of the immune systems in which so much unique individuality resides. His pioneering researches in this important field led him increasingly away from anthropology and he left Oxford to pursue them first with the MRC and later at Newcastle and Birmingham.

Harrison's animal experiments formed the basis for a number of studies of altitude effects on the human biology of Ethiopians. Altitude was chosen for investigation because there can be great variation in it in short geographical distances and therefore in genetically and culturally similar people. This situation occurs in the Simien mountains of northern Ethiopia which are also quite densely populated. The studies showed there were marked differences between highlanders and lowlanders, some of which were due to direct altitudinal effects such as in haematology and lung function measures but that others were due to indirect effects such as the varying economy between highlands and lowlands which affected nutrition and growth.

Harrison also undertook field researches in a number of other situations of human biological interest, in Namibia, Brazil, India, Papua New Guinea and Australia as well as in the UK itself (Harrison and Gibson 1976). They focussed on components of human adaptation, and ranged from investigating the functional significance of genetic polymorphisms to the effects of breast feeding on the body composition of women in Papua New Guinea. Increasingly, attention was given to issues of nutrition and child growth and in collaboration with Gerry Brush (Computing Officer in the department) the changes that occur in the levels of stature and weight variability during growth and in different environments were examined. These changes clearly reveal the effects of environmental quality and of changing capacity for homeostasis.

In the Papua New Guinea researches (a joint British, Australian and PNG IBP project) Harrison collaborated particularly with his colleague Boyce. Following his involvement in numerical taxonomy, Boyce became increasingly involved in researches of human population genetics. He was especially concerned with the genetic structure of populations and examined by matrix analysis, for example, the effects of migration and geographical location on the levels of genetic relatedness between the villages of Karkar Island in PNG which have a circular distribution around a central volcano. By also taking account of village sizes he was able to predict the patterns of gene distributions. Another area in which he was able to use his considerable mathematical talents was in devising quantitative models of people's knowledge of their neighbourhoods to explain marriage patterns between groups of Oxfordshire villages. The wider concerns on population structure are well evidenced in the book *The Structure of Human Populations* (Harrison and Boyce 1972). Boyce also provided great statistical help to researchers in the Institute of Molecular Genetics at the John Radcliffe Hospital (Institute of Medical Research) in their various molecular studies of population relationships (Boyce and Mascie-Taylor 1996).

For some twenty-five years research in biological anthropology at Oxford was run by a trio: Harrison, Boyce and Reynolds. The background and interests of Reynolds were somewhat different from those of the other two. Almost uniquely in the UK he was an academic biological anthropologist who had a first degree in integrated anthropology (from UCL). He was thus much more expert in social dimensions and these have been increasingly recognised as being vital to understanding innumerable problems in human biology. This very special position of Reynolds is well seen in his important books: *The Biology of Human Action* (1976) and *The Biology of Religion* (with Ralph Tanner 1983). His first researches were in primatology but when he came to Oxford, by force of circumstances, he had become involved in studies of child behaviour. It is a remarkable fact that until very recently biological anthropologists had ignored the juvenile phase of human experience – except for child growth which for long has been endlessly investigated. Reynolds continued his child behaviour studies in Oxford, particularly into the nature of autism, but increasingly turned to more broadly ecological issues. He pioneered, for example, the use of heart-rate monitors for examining human activity patterns. In the later stages of his career, however, he returned to primates and especially chimpanzees having become deeply concerned with their survival and conservation. He returned to Uganda where he had undertaken his first work and established the Budongo Forest Project which still prospers. This project focuses on the ecology and behaviour of the forest chimpanzee but it is also much concerned with forest conservation – clearly a prerequisite for chimpanzee survival. Researches are therefore undertaken on the relationships between forest, wildlife and the local human populations (Reynolds 2005).

Further to these largely individual interests, the whole Department of Biological Anthropology was involved from 1965 in a diversity of researches into the total human biology of a group of Oxfordshire villages in the so-called Otmoor region. These studies were not only pursued by almost everyone in the Department, including numerous research assistants employed on outside grants, but also other Oxford colleagues, especially Hiorns of Biomathematics and researchers from Cambridge, particularly John Gibson and Nicholas Mascie-Taylor. The studies began when it was discovered that the ecclesiastical parishes of Otmoor – a rather distinctive region – had excellent church records of baptism, marriage, and burial more or less continuously from the mid-sixteenth century in some cases. These records were copied by Christine Küchemann and Boyce and assembled into individual life histories and re-constituted families. It was thus possible to reconstruct the historical demography of the region and examine the effects of changing environmental and social conditions. Particular attention was given to the analysis of geographical mobility, especially marital mobility because of its profound effect on genetic structure and, from this analysis, predictions could be made about the genetic heterogeneity of present day inhabitants.

The second phase of the study was to test those predictions and for this purpose blood samples were collected from villagers and tested for various genetic polymorphisms, e.g. blood groups, serum proteins and isozymes. In addition, a series of more complex characters were also measured, such as stature, bodyweight, IQ and personality traits. These latter characters have both genetic and environmental determinants to their variation and also may be expected to influence the form of any mobility. The genetic polymorphisms fitted the demographic predictions perfectly, but there was substantial heterogeneity between the villages in the anthropometrics and the psychometrics. IQ was clearly associated with the form of in-migration and even more so with social mobility i.e. the vertical movement between social and occupational groups. Some evidence was also obtained of relationships between various polymorphisms and psychometrics, but these need further investigation.

Since all of the Otmoor researchers were impressed by the goodwill of the villagers and many of them remained interested in the studies, a third stage of investigations was launched. This focussed on states of well-being which are not only of immense practical importance but probably also of great evolutionary significance. Three measures of well-being were employed: (1) medical history, (2) reported sleep patterns especially in sleep duration and sleep latency and (3) levels of adrenaline and noradrenaline output as measured in urine and examined against reported lifestyles and well-being perceptions. Very briefly, in conclusion, sleep duration is related to what people do, sleep latency to how people feel, and adrenaline to arousal levels and how these vary on a daily basis. The results of the whole Otmoor researches are summarised in Harrison (1995) and owe much to the input from Caroline Palmer.

Following the insight gained from the hormone studies of Otmoor, Harrison extended their use to various other population situations. He was also able to add assay of glucocorticoids, and especially cortisol to the assay of catecholamines. A number of interesting findings were made. There are, for example, enormous differences in the levels of excreted adrenaline between populations, with some Polynesian traditional islanders putting out only about a third of that from people in urbanised developed societies. He also showed, with Lincoln Schmitt of the University of Western Australia, that surprisingly, cortisol secretion is less in Aborigines in the towns of northern Australia, where all the problems of violence and alcohol abuse are everywhere evident, than in the outstations where drinking is largely forbidden and violence is rare. Raised cortisol is found in various stress situations such as chronic depressions, and the findings suggest that boredom can be more harmful than social disintegration.

Harrison retired in 1994 after thirty-one years in office. Reynolds in 2001 after twenty-nine years in office and Boyce took on an ever-increasing role in the affairs of St John's College from 1990, after eighteen years in research. These more or less synchronous changes were inevitably going to have a profound effect on the Department. Initially all went well. The Chair in Biological Anthropology became established and Ryk Ward from the University of Utah was appointed. He was very much a geneticist who had undertaken impressive researches on the Yanomama Indians of Venezuela and the West Coast Indians of North America. He also persuaded the University to set up another temporary five-year lecturership in Primate Genetics and Mundy was appointed. Further a long established Departmental Demonstratorship was converted to a Lecturership in Human Ecology to help meet the needs of Human Sciences. The first holder of this post was Stanley Ulijaszek. A further temporary post was created when Boyce's duties at St John's became so time consuming that he largely had to give up departmental responsibilities. This was paid for by the College and Ros Harding, an Australian geneticist, was appointed to it. She is principally needed to meet teaching requirements but she has had a distinguished research career as a mathematical geneticist.

By the time of Ward's appointment human genetics had largely become molecular genetics, a field of enormous medical as well as evolutionary importance. Ward was completely committed to it and began researches in various aspects from establishing the form of genome variety geographically to the molecular genetics of blood pressure variation and the particular susceptibility of peoples of African descent to hypertension when living in modern societies.

Molecular genetic studies, however, are both expensive and require considerable laboratory facilities. Those on the Banbury Road were too modest for Ward's requirements and on his appointment he was given extra laboratory space in the Zoology Department. Even these however became limiting and running a small department on a split site was far from easy. Ward therefore persuaded the University to provide more laboratory space in Zoology and proposed that biological anthropology be moved in its entirety to the Zoology building, which was being extended anyway, and become a sub-department of it. This was due to take place when Ward suddenly died in post. His death was a tragedy for the Institute of Biological Anthropology and the culmination of a series of disasters for the subject. Reynolds had retired and his post 'frozen' at least in part because of the heavy debt the Department had run up; Mundy's temporary lecturership had not been renewed, so despite his excellent researches into primate molecular genetics, especially of marmosets, he left to take up a permanent post in Cambridge. Boyce was mainly involved in affairs at St John's College where he had become Bursar, and Ulijaszek had moved into the Institute of Social and Cultural Anthropology. He came to Oxford as an anthropologist and wanted to remain an anthropologist. He certainly felt that Zoology was the wrong place for biological anthropology in which he recognised the imperative of taking account of social processes in almost every aspect of human biology. He had been very active in research in a number of areas of human ecology: nutrition, growth, body composition and medical anthropology particularly in Papua New Guinea. Even in the short time he has been in Oxford Ulijaszek has published innumerable papers. Ecology is certainly one of the main growth fields in biological anthropology in the rest of the world and cannot be properly pursued without taking account of social dimensions (Ulijaszek and Huss-Ashmore 1997).

A very important role that Oxford Biological Anthropology played in promoting research was by administering the Boise Fund. This fund was established from monies given to the University by Charles Boise, an East African businessman with a deep interest in human evolution especially in Africa. Grants were regularly made from the Fund to support research in this field and mainly to young doctoral research students for field expenses. Support was highly international and many biological anthropologists throughout the world now owe the Boise fund greatly for the help it provided at the outset of their careers.

To summarise, biological anthropology at Oxford has largely been involved throughout its history with present day human populations and other primates, especially chimpanzees. A great diversity of projects has been carried out in many parts of the world but they have tended to focus on elucidating current and recent past population structures. In this various approaches have been used, demographic, genetic, ecological and social and the ways in which these different forces interact to produce a particular overall structure. Focus has also been given to issues of function and especially with the components and processes of adaptation, again from a multidisciplinary perspective of genetics, physiology and behaviour. One of the most significant achievements has been the elucidation of the general relationships between the processes of adaptation and the health and well-being of individuals and populations. It must be clear from this why biological anthropology at Oxford had developed hand in hand with the Human Sciences undergraduate degree, and many of the research students in biological anthropology read Human Sciences for their first degree.

An inevitable consequence of the ever more detailed attention given by biological anthropologists to the detailed structure and function of populations has been the increasing importance of fieldwork in research. In the early days of biological anthropology everywhere, fieldwork mainly involved the collection of materials such as anthropometric measurements or blood samples. This usually required no more than two or three weeks away from the laboratory with little need to be involved with the human subjects, except to obtain informed consent for the study. Very rarely was knowledge of local languages obtained (except for 'please' and 'thank you'!) so investigations could easily move from one part of the world to another. All this has changed. Nowadays, in human ecology long periods of time are required in the field, often for as much as one or two years at a time, and repeat visits and a working knowledge of local languages are essential. Inevitably researchers tend to spend most of their working life with a few peoples and in a particular locality.

The evidence from Oxford is that women are much more willing to undertake these commitments than men. Mention can particularly be given to Catherine Panter-Brick (now Professor at Durham), Melissa Parker, Astier Almedon, Daniella Sief, Mary Ryan and Jo Myers Thompson who undertook long and arduous fieldwork under extremely difficult conditions in recent years. And many others shared to some degree in their experiences.

Another general feature of Oxford biological anthropology and shared with all other similar departments in Britain is in the training of academic teachers and researchers. Here, in contrast to the USA, people tend to move into the subject relatively late. Very few have read for a first degree in any aspect of the subject and most have done their doctorates only in related fields. At Oxford the only established staff member with both a first and second degree in anthropology has ever been Reynolds. All others have first been trained in such fields as genetics, zoology, anatomy, physiology, medicine and geography. From a research point of view there has been some merit in this but the only one with a comprehensive knowledge and total commitment to the whole of anthropology has been Reynolds. This is in marked contrast to the USA where a Ph.D. in anthropology, with substantial course work as well as research, is almost obligatory for a university teaching post in biological anthropology. This ensures a stronger loyalty to anthropology which can be very valuable when conditions become difficult.

With the death of Ward, biological anthropology at Oxford came to a complete halt: practically no staff, no M.Sc. students, no research students, no library and apart from some laboratory space in Zoology, no home!

These dire circumstances arose during a period when the University was undertaking one of its regular internal reviews of the whole of Oxford anthropology. Rightly the committee undertaking the review contained no Oxford anthropologist, just two externals – one in social anthropology and one in biological anthropology. The other members were predominantly biologists but did include an archaeologist and a geographer. Amazingly, it was exactly the same committee which was asked to examine the future of biological anthropology. One might well have thought that with such a remit more expertise in biological anthropology would have been added to the committee. Some external opinion was sought in writing but much of the advice given seems to have been disregarded.

The key recommendations of the committee were that 'the Institute of Biological Anthropology be administratively integrated within the Department of Zoology but that it be allowed to retain its own academic identity', 'the Professorship of Biological Anthropology should be filled within the Department of Zoology' and 'that the field of candidature for the Professorship should not be closely defined but that it should fall within' the fields of 'Molecular human genetics, Human evolution/palaeoanthropology and Primatology'.

The similarity in scope of the field with the syllabus for physical anthropology in the Diploma in Anthropology in 1905 can hardly be missed particularly if one recognises that human molecular genetics in anthropology is in many respects the modern approach to considering the issue of 'race'. One can only conclude that the University sees biological anthropology as of very limited and unchanging scope. Many of the twentieth century developments in human population biology, such as physiological anthropology, auxology, medical and nutritional anthropology and human ecology, have been totally ignored – or at least, not considered worth pursuing. In the world at large they must represent well over half of both the teaching and the research undertaken today in biological anthropology. Time will tell their fate at Oxford.

Postscript

This paper was drafted in 2002. Since then little of consequence appears to have happened. The Chair of Biological Anthropology has been advertised and attracted considerable national and international interest, although without its being filled. Later two senior biological anthropologists were invited to apply but both turned the invitation down. Since then there has been some discussion as to whether the association with Zoology is appropriate but as of the end of 2006 no decision has been taken and no great concern for the situation expressed. Perhaps it will be resurrected but as of now one can only conclude that biological anthropology at Oxford is dead.

Acknowledgements

I wish to express my thanks to my wife, Elizabeth, for preparing this manuscript; to the University Archivist, Mr Simon Bailey who not only made early records available to me but was also able to establish the set up date of 'Supplementary Anthropology'; and to Professor Peter Rivière for many helpful comments.

Academic Staff – Years in Oxford Office

Arthur Thomson Fellow of Hertford College	Professor of Anatomy	1892–1933
Leonard H.D. Buxton Fellow of Exeter College	Reader in Physical Anthropology	1927–39
Sir Wilfred Le Gros Clark Fellow of Hertford College	Professor of Anatomy	1934–62
Joseph S. Weiner	Reader in Physical Anthropology	1945–63
Derek F. Roberts	University Demonstrator in Physical Anthropology	1954–63
Geoffrey A. Harrison Fellow of Linacre College	Reader in Physical Anthropology	1963–76
	Ad hominem Professor in Biological Anthropology	1976–94

John J.T. Owen Fellow of St Cross College	Lecturer in Physical Anthropology	1963–71
Vernon Reynolds Fellow of Magdalen College	Lecturer in Physical Anthropology	1972–96
	Titular Professor in Biological Anthropology	1996–2001
Anthony J. Boyce Tutorial Fellow in Human Sciences, St John's College	Lecturer in Human Biology Ad hominem Reader in Human Population Biology	1972–90 1990–
Ryk H. Ward Fellow of Linacre College	First established Professor of Biological Anthropology	1994–2003
Nicholas Mundy	Temporary Lecturer in Primatology	1996–2002
Stanley J. Ulijaszek Fellow of St Cross College	Lecturer in Human Ecology Transferred to ISCA 2002 and became Titular Professor in 2004	1999–2004
Rosalind Harding Fellow of St John's College	Temporary Lecturer in Human Genetics	2000–06
	Lecturer in Biological Anthropology	2006–

Other Academic Appointments

Departmental Demonstrators: D.F. Roberts, G.A. Harrison, D. Jeffries, R.D. Attenborough, C. Panter-Brick, M. Parker, K. Hill, H. Marriott

Computing Officer: G. Brush

World Colleagues who spent at least one term working in the Institute of Biological Anthropology

Europe
Elena Godina
Lylianne Rosetta

Japan
Tai Takemoto

Australia
David Allbrook
Maciej Henneberg
Les Hyatt
Lenore Manderson
Lincoln Schmitt
Neville White

United States of America
Paul Baker
Cynthia Beall
Bennet Dyke
Gabriel Lasker
Jean MacCluer
William Stini
Alan Swedlund
Brooke Thomas
Al Wessen

Years of Events

1885   Introduction of Anthropology as a Supplementary Subject in Honour School of Natural Science

1905   Establishment of the Diploma in Anthropology

1927   Establishment of the Laboratory of Physical Anthropology

1964   Diploma in Human Biology replaces Diploma in Anthropology

1970   Honour School of Human Sciences

1976   Establishment of Department of Biological Anthropology later to be renamed Institute of Biological Anthropology

1979   M.Sc. in Human Biology established in conjunction with the Diploma

1990   Institute of Biological Anthropology incorporated as a full department within the School of Anthropology

1992   Honour School of Archaeology and Anthropology

Chapter 7

OXFORD ANTHROPOLOGY AS AN EXTRA-CURRICULAR ACTIVITY:

OUAS AND *JASO*

Robert Parkin

Introduction

As well as having a reputation as one of the world's leading universities, Oxford is also famous for its clubs and societies, which between them cover all manner of activities, from sports to dining, hobbies to intellectual interests, publishing college magazines to theatre and music. Indeed, whether considered an extension of high-table dining or a forum for undergraduate extra-curricular activities, the social life of the University has long been one of its great attractions for fellows and students alike. However, controlled ultimately by the Proctor's office in so far as they use the Oxford name, many societies and clubs are really a half-way house between the highly structured regimes of knowledge provided by lectures, tutorials and examinations, and the everyday forms of resistance constituted by the informal activities of students meeting in the college bar or JCR, or just talking in someone's room after Hall over a cup of coffee. Such societies give Oxford its own form of ritual effervescence, separate from student parties and communal living, and constituting semi-institutionalised occasions at which there can be definite expectations of attendance and appropriate behaviour.

The Oxford University Anthropological Society (henceforward OUAS, or simply 'the Society') has enjoyed a longer history than most such societies, its existence running from 1909 to 1989 without more than an occasional brief interruption, followed by a long hibernation until its revival in 2001. It perhaps does not correspond entirely to the above sketch, not least since, for many decades, the graduate monopoly of anthropology in the University gave its membership a somewhat older, not to say more serious and mature profile, making it harder to attract undergraduates. Much later, in 1970, Oxford anthropology also acquired a journal, the *Journal of the Anthropological Society of Oxford* (henceforward *JASO*, the acronym by which it has always so lovingly been known, or simply the *Journal*), also unofficial – not exactly a society so much as a revolving committee recruited through co-option rather than the elections that, at least formally, characterise the Society. I shall discuss the histories of both organisations and their significance for Oxford anthropology in this chapter.[1]

OUAS

Although there was evidently an earlier venture in the 1860s,[2] the Oxford University Anthropological Society I am discussing was launched on 28 January 1909, in the wake of the founding of the first course in anthropology four years earlier. Its immediate inspiration was the success of six lectures on 'Anthropology and the Classics',[3] delivered in Michaelmas Term 1908 by Arthur Evans, Andrew Lang, Gilbert Murray, Frank Jevons, John Myres and William Warde Fowler. Marett, who, as Secretary to the Committee for Anthropology, organised them, then mooted the idea of founding a society in the ensuing Christmas vacation. Marett's involvement with Exeter College ensured that it was in the Old Bursary there that the founding meeting was held, in the presence of over two dozen people. Given Marett's standing as a late evolutionist, it is perhaps hardly surprising that he came together with G.C. Robson of New College, who was interested in local archaeology and had been thinking along similar lines, to initiate this venture – not without some tough negotiations. This connection with archaeology led to the Society running field trips in its early days to sites like White Horse Hill and the Rollright Stones, and a so-called 'field section' was instituted. This was evidently not easy to keep going, and references to it quickly fade, but for many years archaeology was at least as prominent in the Society's profile of itself as anthropology (cf. Hitchcock 1981: 14); indeed, as late as 1933, £5 was given to T.K. Penniman from Society funds for excavations on the Gower Peninsula in South Wales (Receipt in PRM file). The anthropology itself understandably reflected the times in being mostly concerned with folklore, survivals, artefacts and distributions. An early, but undated programme gives as the Society's objectives 'to promote an interest in all its branches by lectures, the reading of papers, discussions and the exhibition of specimens' (PRM file). Not until 1922, when Malinowski gave a paper 'On some customs of the Trobriand Islanders' stressing law and obligation, does anything like a more modern anthropology become evident.[4]

Though of amateur status, the Society tried to appear professional right from the outset. Meetings were advertised in the form of printed invitations, as were subscription reminders, a practice later abandoned on grounds of expense, and proceedings were for a while printed in the *Oxford Magazine*, in which the inaugural meeting was also reported (*Oxford Magazine*, 4 February 1909). This report is scattered with jokey references to Australian ethnography of the time. Thus the meeting itself is described as 'the most inspiring of corrobborees', and the report continues:

> Candidates for initiation presented themselves in large numbers ... to celebrate the mysteries. After blood-brotherhood had been duly drunk in tea, Professor Tylor was elected Honorary President of the Society, that is to say, its *Biamban*, or Superior Being, whilst Mr Henry Balfour consented to serve as Acting President, or Chief of the Medicine-men. A constitution was drawn up, not without heat when the question of the Sexual Taboo came to be raised.

The latter may be a veiled reference to a debate over whether or not women should be admitted: at all events, we are told that 'the matriarchs prevailed'. The report recounts that:

> The subscription was fixed at a moderate figure, and the first half-crown to come in was that of the Honorary President. The following resolutions with regard to the coin in question will be proposed at the next meeting: (1) That it be *churinga*, or sacred, and be considered to contain the External Soul of the Society; (2) that, on the Presidential Birthday, the Chief of the Medicine-men rub it with red ochre; (3) that, during the following week, he observe certain taboos, abstaining from parting his hair with his fingers, from using words containing the syllables 'ty' and 'lor', and from eating witchetty, or other, grub [the singular indicating a possible pun here]; (4) that candidates for the Diploma Examination be permitted, on the day appointed for sending in their names, to press the *churinga* upon their foreheads, the ceremony to be known as that of 'hardening the brain'.

By the end of Hilary Term 1909, the Society had attracted over a hundred members. It met initially in the University Museum, but soon also in other venues such as the Ashmolean Museum, the Pitt Rivers Museum and Barnett House (the latter then in Broad Street, where it accommodated the Institute and the Tylor Library from 1914 to 1921),[5] as well as various colleges. Between 1912 and 1926 there were usually only three actual talks per term, the fourth meeting being devoted to 'Short communications and exhibits'. This practice led to appeals being made for exhibits and photographs to be brought along to meetings, again reflecting the initial inclusion of archaeology and material culture. In these early years too, one talk was often a 'Presidential Address' given by the newly elected president. The practice of not allowing these to be followed by questions and discussion must have made them seem rather like inaugural lectures. Evening meetings were the norm at the start, a pattern maintained up until the Second World War, when the problem of the blackout led to afternoon meetings being held from the middle of 1940. This continued after the War, and in fact it was 1964 before there was a permanent return to evening meetings. Early attendances varied from twelve to 150 or so. Lantern slides were almost routinely shown, a custom that has been maintained, using updated technology – and despite frequent technical problems – throughout the Society's existence. The Society co-operated with similar associations wherever possible in holding joint meetings, and it secured affiliation to the Royal Anthropological Institute in 1909, followed by affiliation to the Folklore Society in 1911. Much later, in 1933, it was to institute a 'folklore group' of its own, which does not appear to have lasted long. In 1913 it sent a delegate to the International Archaeological Congress being held in Ghent, whose warm reception there was duly reported at a later meeting.

Subscriptions were initially half-a-crown, compoundable to a guinea for life membership.[6] The Society never seems to have been very flush with money, and there were repeated attempts to boost income by diplomatically reminding late payers among the existing membership and finding new recruits. Periodic attempts were made to boost membership by attracting not only anthropologists but also those in other disciplines, and early moves were made to recruit an undergraduate secretary in every college in support of the Society. Provision was made from an early stage to recruit associate members who were not members of the University, and in 1923 a class of honorary members was instituted for distinguished scholars, not only anthropologists.

Relatively few of the names of members that have come down to us are well-known today, and the impression of the early history of the Society is very much one of a mass of enthusiasts leavened by the few anthropology and archaeology professionals of the time, not all of whom were invited to speak, let alone to become involved in running the Society. Not until the 1940s does the election of professional anthropologists to the presidency become anything like a regular feature, and even then it is far from being an invariable practice.

Some of the names mentioned prominently in connection with the Society at this time are nonetheless still of interest, such as Henry Balfour, of the Pitt Rivers Museum, and E.T. Leeds, Keeper of Antiquities at the Ashmolean Museum, who served it as treasurer until 1926, and regularly in other capacities into the 1930s. Edward Tylor and Balfour shared the presidency in 1909, to be followed by Marett later in the year (until 1910). It seems that Tylor's role was purely honorary, as well as exceedingly brief, and he soon disappears from the records, apart from a much later attempt, in 1961, to have the Society named after him (see below). A practice soon arose of giving former presidents the status of vice-presidents. The first female president was Beatrice Blackwood in 1931–2, followed by Brenda Seligman in 1934–5. Another woman frequently mentioned in the early records is Miss Freire-Marreco, who, along with another member, Sir Francis Knowles, took the first Diploma examination in 1908 (Brice 1953: 13).

Other names from this period include Sir John Myres, Wykeham Professor of Ancient History, Dudley Buxton, later University Reader in Physical Anthropology, W.E. Le Gros Clark, Professor of Anatomy, Gilbert Murray, expert in Greek literature and religion, L.R. Farnell, Rector of Exeter College, and from 1927 Penniman, Demonstrator and later Curator of the Pitt Rivers Museum.

The first lecture of all, given at the first full meeting of 5 February 1909, was entitled 'Archaeological work on the Zambezi'; it was apparently given by Balfour, the first acting president. The second lecture, on 26 February, was given by A.R. Radcliffe-Brown, then plain Mr Brown, on the Andaman Islands and was entitled 'A pre-totemic religion'. Also in that year, Marett spoke on 'Mr Lang's theory of savage supreme beings', at which 'a number of bull-roarers were exhibited and practical illustrations of their use were given' (minutes, 22 October 1909; cf. Hitchcock 1980a).

Indeed, we learn of this first period: 'There were many lively meetings in the early days, though perhaps none so turbulent as that in 1912 when the long career of Professor Sollas was almost cut short. Ray Lankester threatened him with the alleged rostrocarinate flint implement, whose human origins Sollas had ventured to doubt. Marett wrote a poem to commemorate the occasion'(ibid.: 1).[7] Sollas was Professor of Geology in the University, whose interest in fossils had led him to physical anthropology; he was a keen member of the Society, which he attended into his eighties, and indeed was in the chair as its President at this meeting. In a report of another meeting, this time published in the *Oxford Magazine* for 2 December 1909, at which T.C. Hodson, later professor at Cambridge, talked on the Nagas and a Mrs Scoresby Routledge on the Kikuyu, we are told that the members of the society 'proceeded to give [the speakers] a reminder of the horrors of an Oxford *viva voce*' through their questioning. In yet another report in the *Magazine*, dated 3 February 1910, we are warned that, when it comes to judging speakers' contributions, 'The Society is not one to be trifled with'.

Meetings continued without interruption throughout the First World War. Some of the Society's scarce funds were invested in war loans – so scarce, indeed, that it could not afford a suitable box in which to store the receipts. In 1922 such investments amounted to £56.10s, and the Society was still receiving interest on them (60p for six months) in 1980 (Hitchcock 1981: 15).[8] Some planned speakers were lost to call-up during the hostilities, though not, as far as is known, to the hostilities themselves. This being a mass-conscription war of national survival, it is gratifying to learn that the Society was kept going in these years by its women members – just as they kept the home front going generally (Handwritten note, PRM file).

By 1929 the secretaries had begun producing synopses of talks rather than just minutes of meetings. In that year Evans-Pritchard gave his first talk to the Society on 'Black magic and public opinion in Zande society'. He returned a year later to talk on 'Ethnology of the Anglo-Egyptian Sudan'. The meeting of 30 October that year was held in memoriam for Baldwin Spencer, who had recently died, and was recorded in *The Times*. Among those who spoke at the meeting were his former colleague and correspondent, Sir James Frazer, who was himself to be honoured with a mention on the occasion of his own death in 1941 (minutes, 15 May). The deaths of especially longstanding members were routinely mentioned and a letter of condolence sent to their nearest kin. The meeting of 4 March 1943 was devoted to honouring the recently deceased Marett, a stalwart of the Society from the start. Radcliffe-Brown's death was noted at a 1955 meeting. He had been elected President in 1938, in which capacity he gave an address on 'Myth'.

In general, there were few fundamental changes in the Society's activities or problems in the years immediately following the Second World War. There was the same mix of folklore and archaeology in talks, interspersed with occasional presentations from leading functionalists. Intermittent concerns are evident over both recruiting and retaining membership and raising subscriptions, and the Society seemed perpetually short of money. On 25 February 1953, the five hundredth meeting was held as a commemorative occasion at which many leading figures offered reminiscences, in the main of Oxford anthropology in general rather than the Society in particular; these presentations being privately printed in a pamphlet, *Anthropology at Oxford* (see note 1). This meeting was also recorded in the local press, but the most entertaining account is undoubtedly that which appeared in the *Guardian* for 27 February, which reports some of these reminiscences about early Oxford anthropology:

> Through all the anecdotes told last night – of lecturers who wrote with both hands simultaneously, of those who pulled off a button at the back of their frock coat and referred to them in arguments about vestiges, of others who had 'unquestionable rows' and destroyed each other's malacca cane with arrowheads – shone a tremendous enthusiasm for science. ... There was the wife who exclaimed in a lecture, 'Oh Edward, last time you said that was neolithic.' And the demonstrator who set his beard alight with a bow drill in displaying the ancient art of making fire.[9]

The *Guardian* report[10] indicates that the OUAS had a certain profile in wider intellectual circles. However, the BBC declined an offer to broadcast the reminiscences as a talk or series of talks, their polite rejection being signed, perhaps fortuitously, by one Prudence Smith of the BBC Talks Department.[11] A handwritten note in the PR file containing further anecdotes may also date from this occasion: 'An apocryphal story has it that after one outdoor experiment, a University statute had to be drafted forbidding the throwing of boomerangs in the University Parks.'[12]

The talk given in early 1954 by Thor Heyerdahl on the Kontiki expedition was also reported in the *Oxford Mail* for 28 January; a report containing what appears to be the only published photograph of an actual Society meeting. Royalty visited in 1955, in the person of Prince Peter of Greece and Denmark, speaking not on his customary polyandry but on 'Danish expeditions to Asia'. In general, there is now a noticeable falling away of what might be described as folklore and archaeology as staple topics and a greater shift to the dominance of a recognisable contemporary anthropology, though still with periodic contributions from representatives of allied disciplines, ranging from biological anthropology to ancient history, from literature to philosophy. Europe features quite heavily, though talks are now distinctly anthropological for the most part rather than folkloric in orientation.

In 1961 a proposal to change the name of the Society to the Tylor Society was defeated (it attracted only two votes),[13] as was a motion the following year actually to dissolve it. There were occasional joint meetings with other societies such as the Folklore, Geography and Exploration Societies, and films began to be shown in abundance. Parties are mentioned for the first time in 1964. Two years later there was another motion to dissolve the Society. According to the minutes (for 15 March), 'A few dissident malcontents tried to kill the society, but since it is sustained by ties of deep affection and solidarity and a very long tradition, it seems as if it will survive.' Despite this evidence of trouble, the same set of minutes concluded: 'This was one of the nicest committee meetings yet experienced by the old secretary, since it was particularly small and low in tone.' Whatever the problem was, it cannot have lasted long, since the following year, in a new departure, the Society organised a conference at St John's College on Ecology in the Human Sciences, with Wenner-Gren money, which was deemed a great success. The year 1967 also brought a renewed attempt to attract undergraduates into the Society, and in the following year the question of researching the history of the Society was raised for the first time, followed coincidentally by some external requests to consult the records.

Such sparks of humour as those in the minutes cited above provide occasional relief from records of routine business. In several minutes around this time termly meetings, subscriptions etc. were unaccountably called 'terminal meetings, subscriptions, etc.' The minutes of a meeting held on 2 May 1967 report a fierce tussle between Edwin Ardener and Bernard Fagg, Curator of the Pitt Rivers Museum, *not* to be elected President of the Society. After several rounds of reciprocal proposing and seconding, interspersed with increasingly desperate arguments on both sides, Ardener won (that is to say, did *not* become President). Certainly the burdens of office could get to people: one record of a meeting in 1966 ends with the marginal comment, 'Resignation of secretary: relief, relief, gasp, gasp, what a job!', after which there was apparently a term without any meetings at all – at least, none is recorded. The next secretary had a true sense of the vivid, even the absurd. Thus there are references from his or her[14] pen of a speaker being 'cornered by one of the more persistent questioners' over coffee, to 'usual' problems with projectors, and to 'Mr Ardener [who] – since the lecturer was delayed by the slow service at the Trout Inn – successfully calmed the restive throng with a short film on the Cameroons' (minutes, 22 November 1966). At another meeting in 1966, which incidentally proves that wine was now being served after meetings, a shortage of glasses became apparent. The minutes go on (undated, but the 630th meeting, probably 11 October), 'living up to the high traditions of chivalry cherished by the society, many members were kind enough to accept a large paper cup instead of a tiny wine glass. Indeed, so perfect was the chivalry of some that they claimed to prefer a paper cup.' Such hints of mild inebriation resurface nearly twenty years later, when one speaker ended his talk on alcohol and culture by suggesting 'that the way things are going in England there may come a time when alcohol is prohibited. The audience did not agree with the final conclusion and adjourned for alcoholic drinks at 9.40' (minutes, 28 January 1986).

Among notable speakers in the 1960s were a number from outside anthropology. One was David Attenborough, who in 1961 spoke on 'From New Guinea to Tonga', a talk mainly on cargo cults, but which ended with him declaring a definite preference for Polynesians over Melanesians, whom 'he could only describe ... as neurotic' (minutes, 17 May). Later that same year, Robert Graves spoke on 'The Ambrosia taboo', apparently a talk about magic mushrooms. In 1964 the philosophers Peter Winch and Alexander MacIntyre debated E-P's *Witchcraft, oracles and magic among the Azande* (1937a) before the Society, and in 1969 the Jesuit theologian the Revd F.C. Coplestone defended existentialism to it.

The early 1970s were rather dominated by films, some repeated rather quickly, with often very small attendances, down to seven on one occasion, though others reached above fifty. A 1973 talk on 'Nigerian skin-heads', by which was actually meant masks and helmets, by Keith Nicklin, of the Department of Antiquities in Lagos, evidently caused some confusion for some members of the prospective audience. The minutes (undated, but for the 687th meeting in early 1973) report: 'Amazingly, a large number of the audience believed that they were to hear a discussion of a recent sociological phenomenon, and left.' In MT 1964[15] comes the first mention in the minutes of coffee mornings, which were already producing a profit and were to become a key Society activity, separate from talks, until the Society itself went into hibernation a quarter of a century later. Christmas and other termly parties – no doubt for some revellers also virtually terminal – were the next innovation, another activity that lasted. Indeed, there were parties but no talks of the old style in TT 1976 and again in MT 1977, though student-run seminars make their appearance. In general, the involvement of students in running things becomes more conspicuous from this time on. The Society seems to have been virtually dormant in 1978–9, but in MT of the latter year it was revived by Peter Lienhardt talking on 'Carpet merchants and carpet designers in Iran'. John Beattie was its president from 1979 to 1981 (cf. Coote 1991).

The chief innovation of the early 1980s was a shift of venue to Wolfson College buttery (from MT 1981), where I myself went to many meetings that are still fondly remembered by both myself and many other members of the time. With Mike Hitchcock now the Secretary, many of them were advertised with his cartoons. In 1982, Rosemary Joseph gave a very memorable talk on 'Zulu women's love songs'. The minutes report, with some delicacy (15 June):

The subtlety of the music comes from the harmonics provided by resonances produced by playing the instrument against the naked human breast. Although the speaker herself gave an effective performance on a genuine instrument she declined from providing total ethnographic accuracy. Led by the President [Godfrey Lienhardt] the audience nevertheless showed its appreciation of this musical end to the year by leading a conga around the darkened room![16]

On 27 January 1984, the Society had an intercalary meeting at the Museum of Modern Art in Pembroke Street to celebrate the seventy-fifth anniversary of its founding, viewing some of the RAI's photographic collection. At the end of the year, Rosemary Firth talked about her life as the wife of a famous anthropologist, introducing herself, according to the minutes (4 December), as 'the old lion's mate'. In the New Year, I myself took the minutes for a talk by Joseph Tanega, a first-year graduate student appropriately dressed for the event, on 'Aikido and Misogi: martial arts and purification'. One thing I highlighted in the minutes was the control aimed for in these practices over maintaining the body's centre of gravity. After the meeting Tanega's supervisor, Peter Lienhardt, with whom he evidently already had a pretty jocular relationship, took it upon himself to test this control in his student by giving him a sharp poke in the stomach – entirely without effect. The following term, Trinity 1985, was devoted to a series of lectures on 'Art and anthropology', held jointly with, as well as at, Wolfson; later to become the core of a book of the same title edited by the organisers, Jeremy Coote and Anthony Shelton (Coote and Shelton 1992). One lecture, I recall, was interrupted by a taxi-driver looking for his fare; loud laughter swiftly drove him from the theatre.

A meeting the following year was 'thrown into ... confusion by a possibly unprecedented event in the society's history. Mr Jeremy Coote, the retiring secretary, had failed to produce the minute book. With astonishing presence of mind, Dr Morphy saved the day by blaming the appalling weather' (minutes, 21 October 1986). Coote was quickly replaced – presumably not as a punishment – by Pat Holden, whose words these appear to be. At the very next meeting, on 4 November, there was 'a presentation ... on behalf of Dr Roger and Mrs Tina Just, now in Australia, of a koala bear hand puppet. The chairman received the gift noting that it would provide the society with a much needed totem.' There seems to have been some uncertainty afterwards over what the totem, and indeed the glove puppet, really was, since in the minutes the words 'koala bear' have been crossed out and replaced with 'duckbill platypus'.[17]

Figure 12. Present at the 500th meeting of the Oxford University Anthropological Society on 25 February 1953, from left to right, Wilfred Le Gros Clark, Professor of Human Anatomy; E-P; Sir Alan Pim, former colonial administrator and President of the OUAS, and Tom Penniman, Curator of the Pitt Rivers Museum. Copyright ISCA, University of Oxford.

By now, however, the Society was about to go into a long hibernation, some eighty years after its foundation. There were no meetings in MT 1988, and only films were shown in HT 1989, after which the minute book remains unfilled. The final talk before the long pause that was only to end in 2001 was therefore that given by Malcolm Chapman on 31 May 1988, on the contrasting symbolic geographies of the Lake District and West Cumbria, somewhere in the middle of which he was living at the time. In the intervening twelve or so years the Society was lost from view, especially to the Proctors; even coffee mornings becoming an activity of the Institute, which they remain at the present day.

The reasons for the Society's demise remain somewhat obscure, but it is clear that there was no formal motion to dissolve it. One factor, though, was the less leisurely atmosphere that developed in academia in the 1980s, for both staff and students, with pressures on student funding and tighter completion deadlines, and the first stirrings of that audit culture which has come to dominate so many university activities. In such circumstances, voluntary organisations such as the Society (and, indeed, *JASO*) are likely to be the first to suffer.[18]

The Society was nonetheless eventually revived, in 2001.[19] Its resurrection originated from David Parkin, the current holder of the Chair (as of the time of writing), encouraging students to create their own writing-up and reading group, of a sort he had been familiar with in his previous post in the School of Oriental and African Studies, London University. This led to someone remembering about OUAS, the old minute books being dusted off and the Society being re-launched with Alberto Corsin Jímenez as President, Fernanda Pirie (who had organised the writing-up group) as Secretary and Chris Wingfield as Treasurer. Mette Berg, a later President, was involved in convening a fieldwork seminar. All four were then still students, as have been subsequent officers, and indeed the new-look Society is much more conspicuously student-run than its previous incarnation, with only Marcus Banks involved as the statutory Senior Member. This has meant that it is now at least as much concerned with practical issues of concern to students specifically as with the presentation of purely academic papers. In short, the atmosphere of a gentleman's club with which OUAS started, already pretty redundant when I was a member in the 1980s, has now completely disappeared.

The official re-launch of OUAS took place in October 2001, featuring an address by David Parkin. Peter Rivière and Nick Allen were invited to speak on their experiences as anthropologists in Oxford to inaugurate a new series of talks by leading anthropologists from around the country. After attracting a small but committed initial following, membership rose to forty by HT 2002, by which time the

Society had a newsletter and a website – the latter including a database of research funding, a breakdown of student research projects and a photographic archive. Another initiative was a mentoring programme, in which D.Phil. students undertook to help out with undergraduate summer research projects. The Society also helped organise proper elections to the Joint Consultative Committee, an official University liaison committee linking the Institute with its postgraduate student body. There has also been a shift, whether intentional or by chance, to varying the pattern of the Society's activities each term. Thus some terms have concentrated on social events like parties rather than academic activities. Other terms, however, have concentrated at least as much on the latter, covering not only talks and film screenings, but conferences too, especially the 'Future fields' conference on fieldwork held in December 2003, and another on 'Future pathways of British anthropology' held in May 2005.

JASO

The first issue of *JASO* was published in 1970. It was started by Brian Street and Paul Heelas, because, according to Edwin Ardener, of a felt need for a forum in which critical discussions of structuralism and other current trends and Oxford reactions to them could be aired to a wider world. This was the time of early post-structuralism, in Oxford as in Paris, though Ardener himself cautioned even in 1980 that this, though pervading the contents of the *Journal*, should not be seen as the alpha and omega of Oxford anthropology at the time. In any case, this 'air of "urgent provisionality"', to quote Ardener further (1980: 131), did not last into the second decade, though the engagement with theory has remained, if somewhat diluted. At the very least, the *Journal* has always sought to avoid publishing purely ethnographic articles.

The very first article, 'Meaning for whom?' by Heelas, set the tone for what was to follow. The bulk of the articles and increasing numbers of reviews were written by graduate students, always a minority within their own circles, but famous names contributed right from the start, though rarely trusting the *Journal* with their major intellectual statements. Papers given to OUAS were frequently published later in the *Journal*, as were some of E-P's papers on various proto-anthropologists and the question of primitive mentality, many of the former eventually appearing posthumously in his *The History of Anthropological Thought* (1981).

The reference to the 'Anthropological Society of Oxford' in the title is and always has been fictitious, not to be confused with the Oxford University Anthropology Society. Although there was an early link, with joint sub-committees and so on, the *Journal*'s autonomy was quickly recognised, and the two organisations have long been separate, though with occasional overlapping personnel. One reason for choosing this particular title was evidently to give a suitable status to the *Journal* so that it could benefit from assistance (especially typing) from the University's Club's Office. These benefits, especially typing, were exploited vigorously for reasons of economy in the early days, but they soon caused a strain in *JASO*'s relationship with the Club's Office; when the arrangement ended, the *Journal* retained its fictitious name. Proctor's regulations requiring a 'senior sponsor', Ardener became editorial adviser and acting treasurer. I think it true to say that he never overcame a degree of nervousness about money after the *Journal* became indebted to him to the tune of £80 after a year – mainly, it seems, because of a tendency to give copies away to begin with, a reflection of the desire to put publicity before returns.

Although Ardener occasionally found articles for the *Journal*, he did not exercise any direct editorial control, leaving this to the editors. To begin with these were graduate students, who also collated, though did not type, the early copies in A4 format printed on a stencilling machine. Ardener (ibid.) alludes to the spookiness that must have been experienced by the neighbours of one editor at the tramping of feet above their heads, as the team walked round a table in his room for hours on end collating the stencilled pages into bound-up copies. The editors and, latterly, numerous assistants have always done virtually everything but print the *Journal*, not only editing articles but also producing camera-ready copy for the printers, administering subscriptions and posting it to subscribers. The turnover of editors was understandably high to begin with, stints being ended by fieldwork or job appointments. However, the recruitment of Jonathan Webber in 1979 changed both the format and the turnover of staff. Crown Imperial replaced A4, proper covers were provided (changing colour every issue until the 1996 volume, when we settled on bright red), and the *Journal* was printed properly, though still from typed masters at first; computer-aided production arrived in 1990. For a time the *Journal* was printed in the University's reprographics department, but since 1988 we have used a professional printer, Anthony Rowe of Chippenham, who has served us loyally ever since. Nine 'Occasional Papers', mostly edited volumes, also appeared in these years (from 1982 to 1993).

From Webber's arrival, the *Journal* thus began to enjoy the services of a more stable editorial team. Another leading light after 1986, if more intermittently, was Jeremy Coote, and there were others, especially in the early 1980s, such as Steven Seidenberg, Pat Holden and Elizabeth Munday, followed in the 1990s by Tamara Kohn, Helen Lambert and Chris Holdsworth. I myself joined in 1983 and have remained involved with it ever since. I was recruited by another editor one evening that year over a post-OUAS drink in Wolfson bar. I had just given an article on Munda kinship to Seidenberg (see Parkin 1983), who had waylaid me in the Indian Institute one day to ask me for something for the *Journal*. This rather surprised me, since I thought *JASO* was solely for grown-up students, especially the luminaries of Ardener's circle, who, after all, knew everything. This slight and rather precious piece, my very first publication (Parkin 1983), evidently impressed some secret conclave somewhere that I was fit to be an editor. Ardener himself finally approved me (note I do not say 'approved *of* me', which I was always much less certain about), and I was in. Learning the business came next. This was not exactly a matter of 'sitting next to Nellie', as the phrase has it, but certainly of 'sitting next to Dilley', if the pun be permitted – Roy Dilley, that is, who did a nine-month stint with the *Journal* in 1983 and had been my recruiter in Wolfson bar. This led to many afternoons being lost to our theses and devoted to the work of others, as we sat together trying to work out what on earth some of them were talking about.

Since Webber already had his D.Phil. when I joined, Dilley and Seidenberg went off to jobs, others simply vanished, and Coote and I sooner or later somehow stopped being students, *JASO* eventually ceased to be a purely student-run journal, though students have continued to be involved as assistants, reviews editors, etc. One factor here, as with OUAS, was the pressure on us all, staff and students, with respect to our mainstream activities. One other event, however, namely Ardener's sudden and untimely death in 1987, though it was a great shock to us all, changed little as far as *JASO*'s everyday existence was concerned. Wendy James kindly agreed to take over as the second signatory for the cheques, but otherwise we continued much as before. By this time the Proctors had forgotten all about us, and we about them. We had long since become a kind of adjunct to the Institute, which gave us a room (we have had three different ones altogether) and allowed us to burn electricity without charge, but otherwise left us alone. We have never strictly speaking been the Institute's official journal, but have often been taken as such by the outside world, and there have occasionally been liaison committees and the like, meeting fitfully, if at all, once they have been set up. *JASO* has always maintained a separate budget, carefully managed by Webber, always remaining in the black, though never making a real profit – which was anyway never the aim of the *Journal*. One thing to its credit is that it appears to have inspired or influenced a number of other shoestring journals, both here and abroad.[20] Not a few of its articles have been anthologised and/or translated into other languages.

One further step that the *Journal* might have taken, but never did, was to be published by a leading academic publisher. However, we are currently (at the time of writing) in negotiations, led by Wendy James, with such a publisher, which have reached an advanced stage. These should lead to the creation of a new editorial team, which indeed already exists in embryo, and perhaps a different focus and different title for the new journal (though with the words 'incorporating *JASO*' to provide a sense of continuity). Once this has been done, the present team will bow out (except, possibly, for a background advisory role).

Conclusion

The activities I have described do not of themselves produce degrees or lead to jobs, but they can nonetheless be considered, informally, a part of students' training. The OUAS has always provided a forum in which the Institute's many outside supporters and previous members can gather and discuss anthropology with its existing staff and students, and the latter have been increasingly involved in its actual administration. As for *JASO*, this has been an initial pathway into print for many a student over the years, whether through an article or, more modestly, a book review, as well as providing a number of students with a modicum of experience of academic publishing and a line on a *c.v.* In these respects OUAS and *JASO* both form part of the story of Oxford anthropology, not of course formally or centrally, but as extra-curricular activities providing an adjunct to the Institute's more formal activities.

Notes

1. Although the two organisations have the phrases 'Anthropology Society' and 'Anthropological Society' respectively in their titles and something of a common history, they became and have remained separate soon after the *Journal's* foundation in 1970. The main sources for the OUAS are the minutes and other archives (letters, programmes etc.), now stored in the Pitt Rivers Museum (hereafter 'PRM file', unless otherwise stated); they include *Anthropology at Oxford*, the proceedings of the 500th meeting of the Society, compiled by William Brice and privately printed in 1953 (see Brice (ed.) 1953). Some brief items drawing on the minutes have been published in *JASO* (Coote 1981, 1982, 1985, 1991; Hitchcock 1980a, 1980b, 1981). For the first decade of *JASO* itself, there is an article by Ardener (1980). Otherwise, there are the personal reminiscences of myself and others. I was briefly secretary of the Society in 1985 and have been an editor of *JASO* since 1983.
2. See Anon 1867: 372 (I am grateful to Peter Rivière for providing me with this reference). This early attempt to launch a society does not appear to have lasted long, nor to have had much to do with what we today might recognise as

anthropology, apart from a paper delivered by A.H. Sayce on 'Comparative mythology'. A report in the *Oxford Times* for 20 February 1955 ('Anthropology at Oxford: Society to hold 500th meeting') wrongly traces the foundation of the existing OUAS back to Pitt-Rivers' donation of his weapons collection to the University, which was to become the start of the Museum named after him, in 1884.

3. They were published with the same title that year. See Marett (ed.) 1908.

4. Space does not permit me to list more than the occasional paper presented to the Society. Instead the focus will be on its overall historical trajectory, interspersed with flashes of often unintended humour arising out of the minutes and other documents. Similar remarks apply to my treatment of JASO later in the chapter (which, since I have long been an editor, inevitably involves a bit of self-mockery too). No doubt both organisations deserve their own, more extended and more serious histories in the future, but, in being extra-curricular, the Society in particular was always intended as a forum for entertainment and as a social event as much as for enlightenment and education. I intend no disrespect to either organisation, nor to those who have given, often generously, of their time in running them.

5. Peter Rivière, personal communication.

6. Respectively 12.5p and £1.05p in decimal currency. The subscription was raised to five shillings (25p) in 1927, and to 7/6d (37.5p) in 1950 (handwritten notes, anon., PRM file).

7. A handwritten account of this exchange in the OUAS file in the Pitt Rivers Museum has Sollas say: 'I ask any impartial person, to what possible human use could that lump of road metal be put?', to which Lankester allegedly replied: 'Well, I could kill you with it'. Sollas continued to complain about 'scrapers that will not scrape, and borers that will not bore', to which Lankester, increasingly angry, retorted: 'You, at any rate, are one of those borers that can bore'.

8. In the interwar years subscriptions yielded from £10 to £20 a year, and the bank balance was normally between £20 and £40 (PRM file).

9. Both of the latter anecdotes refer to Tylor, as does that about the buttons (see Brice 1953: 6, 10). The ambidextrous lecturer was Arthur Thomson, the physical anthropologist (ibid.: 13). One of the contributors of reminiscences, Gilbert Murray, objected mildly to his words to the meeting being published (letter to W.C. Brice, OUAS Secretary, 19 February 1953).

10. Written by Anthony Smith, a former zoology student at Balliol (PR file, letter, W.E. Cockburn, *Guardian* News Editor, to Brice, 19 February 1953).

11. Letter to Brice, 13 March 1953.

12. 'The beginning of the study of anthropology at Oxford', handwritten ms by Brice, date uncertain.

13. The proposer evidently felt that the Society had drifted away from its earlier moorings, when it had made strong attempts to attract undergraduate members. Now, 'It had no base and was all apex' (minutes, 7 June).

14. The anonymity of officials here and elsewhere is due less to delicacy on my part than to a failure to identify them consistently in the Society's records after 1953 (there is a complete list of officers up to that date in the PRM file).

15. MT = Michaelmas Term (autumn); HT = Hilary Term (winter to spring); TT= Trinity Term (summer).

16. I would particularly like to thank Rosemary for allowing me to cite this passage from the minutes.

17. Howard Morphy informed me after the conference presentation that it was indeed the latter, and that he had himself corrected Jeremy Coote's error in the minutes.

18. This paragraph has been informed by discussion after the original paper at the conference, involving especially Helen Lambert and Marcus Banks.
19. I am grateful to personal communications from Alberto Corsin Jímenez, Fernanda Pirie and Istvan Praed for information concerning the Society's resurrection and current activities.
20. It would be invidious to suggest examples, all of which no doubt have their own ideas about their origins and identities.

Chapter 8

OXFORD ANTHROPOLOGY SINCE 1970:

THROUGH SCHISMOGENESIS TO A NEW TESTAMENT

Jonathan Benthall

Many readers of this book will know much more than I do about my subject. I feel like someone arriving at the Newcastle miners' gala with a small scuttle of charcoal. My own first experience of Oxford anthropology was the 1973 ASA decennial conference. Organising some lectures at the ever up-to-the-minute Institute of Contemporary Arts in London, I had disclosed to Edwin Ardener an interest in structuralism. 'Oh, I think we have something rather newer to offer you' he said temptingly, inviting me to Oxford for a day. But this was a national occasion: Ronald Frankenberg was explaining that Lévi-Strauss's structural analysis and Mao Zedong's theory of contradictions could be easily mapped on to each other; and critics of Maurice Godelier drew diagrams explaining the difference between the base, the mode of production and the *social* mode of production. (All this had not much to do with Oxford *per se*, any more than was to be the ASA decennial twenty years later, also hosted by this University but dominated by a post-modern luxuriance that again had little to do with *Oxford* Oxford.) For a first impression of Oxford anthropology unmediated by such extraneous concerns I turn to the reminiscences of a friend.

In the autumn of 1971 an observant and energetic 23-year-old from the American South came over on the QE2 with other Rhodes Scholars. The Warden of Rhodes House met them at Southampton and joined them on the coach to Oxford. When they glimpsed the famous spires, they cheered and clapped. Matt had a first degree from Yale in French with special reference to African writers, and wanted to read comparative literature. The Oxford authorities turned him down for this, and naturally recommended anthropology instead. From day one, Matt became aware of tensions in the Institute of Social Anthropology. He was particularly struck by the importance attached to pub life, and felt some cross-generational sympathy, as a fellow-outsider, with the recently appointed professor from LSE, Maurice Freedman, who rarely came to the pubs. John Beattie had recently accepted a chair in Leiden, in lieu of the chair vacated by Evans-Pritchard for which he had been passed over, and the principal poles of attraction for the Institute's loyalties were Edwin Ardener and Rodney Needham. Matt met E-P once, but by then he was a somewhat ghostly presence. As a West Africanist, Matt started as Ardener's student, but this country boy from Georgia found Ardener's elliptical style hard to cope with. His first months as a graduate student in Oxford were difficult, and not only because of the new experience of feeding gas meters with shillings in an icy bedroom. 'Matt – that's a cowboy name', one of the dons told him. In the course of his academic tribulations, Matt switched supervisors to Needham, who was not only lucid in speech but also positively liked Americans. (No special significance should of course be attached to the switch of supervisors, which happens frequently for a variety of reasons.)

After four years, Matt was awarded his D.Phil., on the Mandinka of Senegal, and he soon published a monograph in the Spindler series (Schaffer and Cooper 1987). He became a successful international business consultant, but his Oxford years were deeply formative and recently he returned in his spare time to academic work with some publications on the politically charged issue of Afro-American cultural and lexical continuities (Schaffer 2003, 2005). He remained a friend of Needham's until they both died in December 2006.

Matt Schaffer's encounter with the Old World over thirty years ago brings out what has been described as the 'internecine solidarity' of Oxford anthropology.[1] E-P – affectionately known to insiders by his initials, as is the custom in Britain for admired company bosses – gave inspired leadership but left no anointed heir. Social anthropology enjoys a tradition comparable to the Christian apostolic succession or the *isnād*, the chain of reporting whereby sayings and actions of the Prophet Muhammad were authenticated. 'From the very earliest days that I can recall', Needham has said in a video interview, 'the Institute was riven subliminally ... on grounds of religious commitment, of sexual predilection and of lifestyle – by which I mean drinking.' (Needham 1979). Yet, he went on, a true *conscience collective* had existed during the early 1950s. 'We all focused intensely on Evans-Pritchard.'

Whether or not it is healthy for charismatic leaders to anoint their heirs is debatable, and their relationship with students and disciples is always complex. The narrator of one of Kazuo Ishiguro's novels has put it well:

> One supposes all groups of pupils tend to have a leader figure – someone whose abilities the teacher has singled out as an example for the others to follow. And it is this leading pupil, by virtue of his having the strongest grasp of his teacher's ideas, who will tend to function, as did Sasaki, as the main interpreter of those ideas to the less able or less experienced pupils. But by the same token, it is this same leading pupil who is most likely to see shortcomings in the teacher's work, or else develop views of his own divergent from those of his teacher. In theory, of course, a good teacher should accept this tendency – indeed, welcome it as a sign that he has brought his pupil to a point of maturity. In practice, however, the emotions involved can be quite complicated. Sometimes, when one has nurtured a gifted pupil long and hard, it is difficult to see any such maturing of talent as anything other than treachery, and some regrettable situations are apt to arise (Ishiguro 1986: 141–2).

The charge has sometimes been laid against E-P that he played off one junior figure against another. Historians of anthropology will no doubt contrast E-P's personal intellectual legacy with that of two figures of similar stature: Malinowski's – which was handled with corporate professionalism by LSE under the aegis of Firth – and that of Lévi-Strauss, beneath whose shadow during his long and productive retirement the next generations of French anthropology have lost most of the éclat that the discipline used to enjoy at the heart of Parisian intellectual life.

An element of theatre, or as Wendy James might prefer to say 'choreography' (James 2003), is no doubt healthy to academic life. Leach knew this well, and cultivated a sparring relationship with Fortes and Goody at Cambridge, and a jocularly oedipal relationship with his close friend Firth. I am reminded of an interview given by the great soprano, Renata Tebaldi, to *Le Quotidien de Paris* in 1986:

> Two clans were formed, the callassiens and the tebaldistes. It was a dream for journalists and for us too, because hardly a day went by without an article appearing about one of us. In the theatre as in sport, people love rivalries. Callas against Tebaldi, it was rather like Maradona against Platini.[2]

However, it seems that at Oxford between 1970 and 1987 the atmosphere degenerated. What should have been the sunlit, exhilarating spirit of Degas' *Young Spartans Exercise at Wrestling* became more like Goya's so-called black painting, *Cats Fighting*.

It seems that there was more to unite than to divide the *conscience collective* intellectually. The differences were more temperamental. I owe both Needham and Ardener a considerable private debt; personal in the case of Needham as one still went to consult him like the oracle at Delphi; institutional in the case of Ardener because he was a doughty champion in the 1980s of struggles to secure for anthropology its rightful place in our country's public life. Oxford anthropology was never gripped by the Marxism that during the early part of our set period held many British contemporaries in thrall, for which reason it has sometimes been pilloried for the intellectual crime of 'idealism'.[3] Nor during the latter period did it go in for deconstruction and post-modernism. Needham and Ardener shared a reverence for Mauss and the *Année Sociologique*, as well as a deep interest in language, and more particularly in holding standard expressions up to the light: 'kinship' or 'belief' in the former case, 'ethnicity' or 'remote areas' in the latter.

Each had something of the poet in him. Needham's modulated prose cadences could sometimes read like verse from *Four Quartets*, as in the following passage where he contrasts the moral benefit to be derived from access to forms of art:

> *Perhaps these advantages really can be secured –*
> *Even on rather contrived conditions*
> *And in a logical scheme that is tautological ...*

with the various inner changes made possible by the 'multifarious testimonies of ethnography':

> *Yet none of this will automatically secure*
> *Within us an altered commitment of the kind*
> *That we regard as moral* (Needham 1985: 42).

Maybe there is a valid analogy with T.S. Eliot's attachment to spiritual exercises, classicism and hallowed institutions.

Ardener had a gift for the sudden arresting simile:

> ...We all have to guard against *over-determining a distinction in our own culture, objectifying it through new data, and then receiving it back, no longer able to recognize our own artefact.* 'Behaviour' is such a case: we may clutch it as those experimental monkey infants clutch their mothers made of wire, and receive precious little nourishment (Ardener, E. 1989: 108, his italics).

Viewed less flatteringly, both could appear mannered, but in different ways. Needham had the military precision of a former

Gurkha officer coupled with a slightly archaic vocabulary. He might for instance, in giving a job-reference, write of the candidate's 'address', meaning not where he lived, but his physical bearing. Ardener must have been the last man in England to wear galoshes to keep out the rain – at least off the theatrical stage; and he is looked back on in Robin Fox's memoir as 'the Max Beerbohm ultimate-Oxford character' (Fox 2004: 454).

Both were devoted to their colleges, both attracted cultic loyalty from students. After the administrative crisis[4] that caused Needham to withdraw from the Institute's life in 1977, he retreated to an eyrie in All Souls whence he continued to exert considerable intellectual leadership until his retirement from the Chair in 1990. On the death of Leach in 1989, Lévi-Strauss found it natural to write his letter of condolence, on behalf of French anthropology, to Needham – despite a long history of Anglo-French misunderstandings on structuralist matters – since for him Needham was now the leader of the profession of social anthropology in Britain (Lévi-Strauss 1989). As for Ardener, his Socratic demeanour prompted a witty compliment from Jack Goody after the ASA dinner during its centennial conference in Cambridge in 1983: he presented Ardener with a large stalk of hemlock. It is a sad irony that Ardener was to die a few months before his sixtieth birthday.

Up until the early 1970s, Oxford anthropology was a convivial society but also a closed shop. Geographically the staff's interests remained focused almost entirely on Africa and India. Despite its considerable scholarly achievements, was it not held back institutionally by memories of its charismatic past and by the ongoing trauma of schismogenesis? Its maturing into a department more fully able to realise its full potential was delayed by leadership problems, or perhaps more accurately by the absence of a felt need for leadership.

The first opportunity for an outsider to break the mould came in 1970 with the appointment of Maurice Freedman from LSE, a pioneer of the social anthropology of China, to the Chair and its accompanying All Souls Fellowship. Ardener had also graduated from LSE but from 1963 had become an Oxford caryatid. It was not that Freedman did not take to Oxford: he had fallen in love with the place earlier and especially with All Souls. He remained close in spirit to LSE, where he was a governor and held in great respect, and maybe the old antagonism between E-P and LSE anthropologists had something to do with the matter, for LSE seems to have been Oxford anthropology's principal 'Other', more than Cambridge. There were no academic or ideological differences between Freedman and the Institute. With a first degree in English he had a fastidious style of writing that might well have appealed to literary sophisticates such as Godfrey Lienhardt.

What seems to have been resented was simply that he was an outsider.
One of the Oxford dons is reported by his widow Dr Judith Freedman[5]
to have said to her, 'I will teach him how to behave in Oxford', and to
have said to Maurice himself, 'Didn't you think of asking us if we
wanted you to come?' This might seem to be unreliable evidence, but a
much later arrival at the Institute has told me that an attempt to
introduce reforms was met with the objection that this was not 'the
Oxford way'.

*Figure 13. Maurice Freedman, Professor of Social Anthropology 1970–5.
Copyright All Souls College, Oxford.*

Freedman had been an Acting Major in India at the age of 23 and had faced down the LSE student disturbances. An obituarist, Arthur Wolf, recalled him as a 'vigorous, strong-minded man who inspired those qualities in others' (Wolf 1975). But his leadership qualities could not cope with Oxford anthropology, including its administrative back-up, and he remained semi-detached. It has been suggested that he seemed to treat the Oxford postgraduates in anthropology as if they were LSE undergraduates. Perhaps a clue is to be found in something his widow has told me: he had been the adored only son of a Jewish East End family and his life had been somewhat sheltered. He died of a heart attack in July 1975 in his flat in London, after only four years in the Oxford Chair and aged only fifty-four.

Needham had been considered for the Chair in 1970 and was now appointed in succession to Freedman. Whatever his intellectual superiority, he was not a new broom and schismogenesis was the order of the day until the late 1980s. This at least was the impression formed by many outside observers of the Oxford scene. The role of the journalists who stoked the Callas – Tebaldi operatic rivalry was taken by anthropological gossips. However, several insider accounts of these years diverge from this outside view. Many students and staff carried on working quietly and productively – the personal tensions described above passing them by.

After losing Peter Lienhardt in 1986 and Edwin Ardener in 1987, and under requirements for retrenchment across anthropology as a whole, plans were made to reorganise the discipline under one School. A number of new staff from outside, of whom Marcus Banks was the first, were appointed. Since then the Institute has had a good record in selecting from a wide variety of talent, thus compensating for the natural inclination of academics to reproduce themselves through webs of patronage.

In 1990, on the retirement of Needham, the chair of social anthropology was advertised and another outside candidate was appointed, John Davis from the University of Kent, another LSE alumnus (though with a first degree from Oxford) with a fine reputation not only as a scholar but also as a technical innovator. Much had changed in Oxford anthropology during the past fifteen years, but Needham's circle of influence radiating from All Souls had stood to 51 Banbury Road as the Papacy of Avignon once stood to Rome. Here was a new new broom.

Fortunately for the University as a whole but unfortunately for the Institute, Davis proved such a personal success that he was translated rapidly, before his impact on the Institute had time to crystallise, into the Wardenship of All Souls, an academic post no less eminent in reality than its other-worldly title suggests. The present incumbent of the chair, David Parkin, also an outside candidate and a vigorous innovator, succeeded him in 1996. But looking back over the period since 1970, it seems that Providence ordained two setbacks in the process of what one can only call modernisation – one caused by the tragedy of a premature death, the other by the honour accorded to anthropology in the advancement to high office of one of its leaders.

I said modernisation – and as so often when one uses that word, a measure of homogenisation is implied. British social anthropology in 1970 remained fairly sure of its identity, and within it the major departments each had a clear distinguishing character, usually deriving from a charismatic leader or leaders. Today, American anthropology is no longer regarded as a separate sub-species, and with one or two exceptions departmental identity seems more elusive. I have not been commissioned to sketch a general study of British social anthropology since 1970, besides which I have already put in my two cents' worth on those intellectually turbulent years in my introduction to an anthology, *The Best of Anthropology Today* (Benthall 2002). I must try to delineate some themes specific to Oxford and I have identified five. The regional specialisation of the Institute has almost disappeared, though it is worth remembering that Oxford had pioneered the social anthropology of Europe with the work of Julian Pitt-Rivers and John Campbell. My search for visible constellations will be unfair to those individual stars – such as Robert Barnes or Paul Dresch – who do not fit into them so easily but have shone with no less brilliance, reminding us that still to this day social anthropology, at least in Britain, remains a profoundly individualist discipline, to which the contribution of teamed projects has so far been intellectually marginal. Indeed, if a constellation is a kind of glorified pigeonhole, a scholar such as Nicholas Allen – combining in his early career such diverse influences as traditional Classics, mountaineering and medicine, and dwelling latterly on the ambitious but still unfashionable field of Indo-European cultural comparativism (Allen 2003) – represents the kind of creative idiosyncrasy that is growing rarer in our increasingly bureaucratised universities.

The first theme, language and meaning, I have touched on already. This is tricky because a concern with semantics has been one of the guiding themes of anthropology in general during the period since 1980, rather than specific to Oxford. The connection between social-cultural anthropology and the study and philosophy of language has been extensive and sustained – despite tactical moves to reduce it, whether by the proponents of kinesics, proxemics and paralinguistics, and later by Bourdieu with his concept of *habitus*, by Ernest Gellner with his sledgehammer approach to linguistic subtleties, or by Maurice Bloch with his arguments against an over-linguistic model of culture.

It is hard to see any close connections between Oxford social anthropology and its elder sister, Oxford linguistic or ordinary language philosophy, considering their physical proximity; but Wittgenstein is a frequent reference point. Ardener's essay 'Social Anthropology and Language' (Ardener 1989 [1971]) argued that it was E-P who did most to bring British anthropology closer to linguistics. I am hesitant about identifying language as a specifically Oxford theme since 1970, but it is irresistible to point out the similarities between, for instance, Needham's analysis of polythetic classification (Needham 1975) and Ardener's concept of 'semantic density' – a 'statistical feature, at the point where definition and measurement intersect and collapse together' (Ardener 1989 [1987]: 211). However, despite the present professor David Parkin's personal interest in linguistics it does not seem any more to be a central focus.

With my second theme, religion, I am on surer ground, because any broad-brush history of mid-twentieth-century anthropology worldwide will surely hail *Nuer Religion* and *Divinity and Experience* as among the most distinctive contributions of the British school (Evans-Pritchard 1956, Lienhardt 1961). Here I take the liberty of trespassing a little outside my period to take issue with Ahmed Al-Shahi's attempt to inter the question of the connection between Catholicism and Oxford anthropology (Al-Shahi 1999). Al-Shahi has no trouble in showing that Adam Kuper's original *aperçu* was a little careless (Kuper 1973: 158). But the salience in the saga of Oxford anthropology of the names of E-P, the Lienhardts, Douglas and Pocock, and of the *Blackfriars* journal – all contributors to the shift away from functionalism, to which names one may add that of Victor Turner, who was not an Oxford product but had considerable influence there – surely merits more consideration than Al-Shahi's reliance on Godfrey Lienhardt's supposedly 'authoritative' view that there was no Oxford Catholic nexus. We have on record the statements of Needham and Raymond Firth that religious differences were indeed an issue in mid-twentieth-century British anthropology – 'Different assumptions about the validity of religion and the nature of religious knowledge have divided anthropology more deeply than have any other positions, even though in public we have been politely reticent about this' (Firth 1981: 582).[6] Surely the notoriously sensitive questions of religious belief and affiliation call for triangulation rather than mere acceptance of the assertions of one of the protagonists.

It is easy to see the interest in religion prolonged throughout the post-1970 period. A number of well-known specialists in this broad field such as Jonathan Webber, Paul Heelas and Fiona Bowie completed their doctorates at the Institute. Wendy James has been the most loyal heir to the E-P legacy, though her monograph on the rituals and concepts of personhood of the Uduk of Sudan (James 1988), embodies a sophisticated approach to regional politics, denied to her Master, that reminds us that she was one of the contributors to Talal Asad's groundbreaking *Anthropology and the Colonial Encounter* (Asad 1973).

The issue of personhood – fundamental to anthropology in the older, pre-Victorian sense as an aspect of theology – is explored in two articles by Peter Rivière, each of which anticipated important research trends. In his Malinowski Lecture on the *couvade*, analysed as a form of spiritual parallel to physical birth, he explored what he saw as an almost universal problem concerning the nature of humanity, the conceptualised dualism of body and soul (Rivière 1974). In a later article (1985), on the Warnock Report on human fertilisation and embryology, he showed how science alone can give no satisfactory answer to the ethical problem of the ontogenetic point – somewhere between conception and a newborn's acquisition of multiple social relations – at which personhood is assumed. A similar issue comes up in Silvia Rodgers' article in *RAIN*, based on her Oxford doctoral thesis supervised by Needham, about the launching of British naval ships (Rodgers 1984) – a ceremony which has a Christian component about which the Anglican clergy feel so uneasy that they invariably field a humble local vicar rather than a church dignitary. Why? Because the quasi-baptismal ceremony assumes that naval ships have souls as well as bodies, just as we do.

My third theme is the anthropology of suffering. In the United Kingdom one might expect LSE or Manchester or one of the newer universities to have pioneered this field, but it was Oxford. Davis had indeed been a former student of Lucy Mair, the pioneer of applied anthropology at LSE, but since her retirement the anthropology department at LSE had been so confident in its academic legacy from Malinowski, Firth and Schapera that for many years it more or less insulated itself from the wider meliorist tradition of the LSE founders, represented within the department only by Peter Loizos and one or two others.

Soon after his arrival in Oxford, Davis took the phrase for the title of his Elizabeth Colson Lecture (Davis 1992). He distinguished the 'comfortable' anthropology of 'maintenance', which documents social structure and function, from the anthropology of 'repair' concerned with issues of policy and intervention. He quoted Colson's aphorism 'Social organization hurts' – which is known in another intellectual tradition as 'structural violence' (Galtung 1969). But Davis went on to make the crucial point that whereas 'soft' anthropology was supported by hard money, 'hard' anthropology was supported only by 'soft' money – i.e. *ad hoc* grants and contracts.

   The driving force behind this tradition at Oxford was certainly the Refugee Studies Programme (RSP), later redesignated Refugee Studies Centre, founded in 1982 as part of Queen Elizabeth House, the University's centre for development studies. Its founder, Barbara Harrell-Bond, was not the first British anthropologist to concern herself with refugee and forced migration issues – Peter Loizos and Frances D'Souza were among those who preceded her (Loizos 1981, Minority Rights Group 1980) – but her achievement in building up the RSP, until her retirement in 1996, into a substantial centre of research, publication and teaching was a notable and gutsy bootstrap operation. Specially imaginative was the way in which the original focus on forced migration radiated out, in true holistic fashion, toward kindred subject-areas. For instance, Harrell-Bond's own book *Imposing Aid* (Harrell-Bond 1986) was one of the first major studies of international NGOs, on which there is now a mass of published research (see also Harrell-Bond and Voutira 1992). A number of influential Oxford-trained anthropologists such as Renée Hirschon and Alex de Waal have devoted their efforts to the anthropology of suffering, while Wendy James's regional interest in Sudanese minorities has also introduced a harsh focus. Kirsten Hastrup took up Davis's metaphor of hard and soft in an important article 'Hunger and the hardness of facts' (Hastrup 1993).
   It might seem that the tables have been turned financially with the recently founded large-scale ESRC Centre on Migration, Policy and Society (COMPAS), interdisciplinary but structurally part of ISCA and directed by Steven Vertovec, an Oxford trained anthropologist. This was set up to study the conditions surrounding migration in areas of origin, transit and destination, and is designed to complement and co-operate with the Refugee Studies Centre but with a much larger budget – about to be expanded to a global scale with the help of private funding, and including all forms of migration. Has 'hard' anthropology now earned 'hard' funding? Not quite, because even the most generous ESRC or private funding is not the same as an endowed Chair, and remains generously soft, like a soft sofa one has difficulty in getting up from. However, what Davis called the 'anthropology of repair' has established its credentials convincingly in the world of aid and development, and it is likely that research funding of this type will continue to be available from many sources. We may expect Oxford, having led the way (at least in Britain) in exploring the anthropology of suffering, to maintain its high standards and in particular to guard against the risk that applied anthropology always runs of becoming opportunistic and/or servile to powerful interests.
   My fourth theme is gender and the social anthropology of women. So much folklore circulates on the marginalisation of women,

especially wives, in traditional Oxford that, again, one might expect the development of gender studies within British anthropology to have been pioneered elsewhere. But maybe it was the spur of marginalisation that provoked Oxford anthropology – itself marginal in the university – to be a leader in innovative thinking about gender at a time when otherwise the United States would have had it nearly all to themselves. The foundations were laid in Oxford at about the same time as parallel developments in the United States (Rosaldo and Lamphere 1974, Reiter 1975). True, it was not within the Institute that the innovation happened, though Edwin clearly deserves a large share of credit with Shirley Ardener, the founder director of the Centre for Cross-Cultural Research on Women, formally recognised, also under the auspices of Queen Elizabeth House, in 1984 after twelve years' existence as an informal seminar. In 2002 it was renamed the International Gender Studies Centre. Though others[7] played their part, the main inspiration and drive behind the Centre were Shirley's. The key insight of the Ardeners, borrowing the term from Charlotte Hardman, was to recognise the mutedness of subordinated groups, that is to say the constraints on their self-expression imposed by the 'deafness' of dominant groups (Ardener S., ed. 1975, ed. 1981, 2005).

The foundation of a women's group was the logical response. Criticised from the beginning on the grounds that such a group was discriminatory, its founders defended the concept saying that when their seminars were thrown open to men, they dominated question time. They were deliberately challenging the hallowed but macho LSE style of 'seminar culture', which is like a sword dance. They had to put up with coarse disparagement as a knitting circle, or with subtler stereotyping as controversial feminists (Ardener, S. 1985). It has earned considerable praise both within and outside Oxford University, the patronage of no less than Aung Sang Suu Kyi, and a State honour for Shirley Ardener; its memorial lectures have been well attended; but this support was never translated into money.

The difficulty for the University is clearly that in the thirty plus years since the women's seminar started, the discovery of gender as an analytical tool has transformed anthropology, for gender differences were exposed as probably the most deep-seated of all cultural differences. The mainstream has absorbed much of the feminist view, just as it has absorbed other intellectual movements. However, may it not be an error to hold that 'we are all feminists now', while some such as Oxford's Laura Rival still believe that gender is at the core of theoretical development in anthropology? There is surely space for at least one focal point in British anthropology for an explicitly gendered approach, especially because questions of gender, particularly reproductive health rights, are among the most difficult in the field of development studies. (In fact, the Fertility and Reproductive Studies Group, founded by Soraya Tremayne with David Parkin's support in 1998, is now integrated with ISCA though still partly dependent on 'soft' funding – another example of innovation at Oxford beginning from the margins.) There is no other centre for gendered anthropology in the United Kingdom than Oxford, and may it be that prejudice on the part of funding agencies is partly responsible for the lack of financial support? It seems to be a sociological fact that when professions in our society are taken over by women, they lose prestige. Is anthropology able to resist this defensive strategy by male holders of power?

My fifth theme is visual anthropology, which was spearheaded in Oxford by Marcus Banks and Howard Morphy (Morphy and Banks 1997). The definition of visual anthropology continues to be contested. During the period under review, the RAI played a leading role both in promoting anthropological film through its Film Committee – two of whose chairmen have been Banks and André Singer, the latter one of E-P's last students who became a prominent ethnographic film-maker – and in the interpretation of anthropological photo-archives through its Photographic Committee, where a key link with Oxford was Elizabeth Edwards of the Pitt Rivers Museum, who edited an important collection of studies of the RAI's photographic collection (Edwards 1992). The two committees had little in common at that time since the former was primarily interested in encouraging new documentary film-making, with ancillary services such as a film and video lending library, whereas the latter was mainly concerned with historical analysis. Morphy's distinction in the anthropology of visual art and Banks's commitment to film, together with the appointment of Michael O'Hanlon as Director of the Pitt Rivers Museum in 1998 and Edwards's continued publishing record (Edwards 2001), have resulted in Oxford's now having an enviably broad base in visual anthropology, with potential for sustained cooperation between the Institute and the Museum.

The repatriation of anthropological photographs to indigenous groups as a vital part of their heritage has become a major part of the Museum's work. For instance, the Tibetan project, led by Clare Harris, is based on extremely rich holdings, in the Pitt Rivers Museum and the British Museum, of photographs taken by British political officials between 1920 and 1950. In many instances, these images are the only surviving record of aspects of Tibetan culture destroyed during the Chinese invasion and later the Cultural Revolution. Use of up-to-date computer technology makes these images available for Tibetans and scholars of Tibet, wherever they may be.

If one question may be asked of Oxford's otherwise outstanding record during my set period, it relates to the Human Sciences degree, which admitted its first undergraduates in 1970. It is true that entry is highly competitive, which is an objective sign of real success, though it has remained small, with an intake of only about forty students per year. The students are reportedly of high quality and it has produced many outstanding graduates.

However, it does seem an opportunity not sufficiently grasped that one of the world's great universities should have devised such a course without making an earlier effort at the highest level to bridge the intellectual, more precisely the epistemological divide between the various components of the course, rather than leaving this task to novices. It has been like a pentathlon contest with separate coaches for wrestling, disc-throwing and the rest, whereas a true human sciences course would be an integrated game like Ultimate Frisbee.[8] After years of trying to facilitate the melding of social and biological anthropology within the RAI during the 1980s and 1990s, I came to the conclusion that there were deep epistemological differences that only a few anthropologists, the Ingolds and Littlewoods and Foleys, were really interested in bridging. Though the best students must surely confront these differences squarely, I can see no sense of an agonized struggle in the present prospectus for the Oxford human sciences course. By contrast, the International Baccalaureate's Diploma Programme – obviously at a more elementary level, since it is intended for sixteen to nineteen year olds – has made its 'theory of knowledge' module the obligatory core of its entire syllabus.

I am fortified in this criticism by reading the following in Robin Fox's memoir of Oxford in the early 1970s (he refers to himself in the third person like Julius Caesar):

> He was asked to help with the new Human Sciences degree course ... that was being introduced. This was intended as a fusion of social science and biology, but no one seemed clear how to do that, except to have students dabble across the board. There had been talk of having some 'integrative seminars' and this is where he was supposed to come in. But then it was decided to 'let integration occur in the natural course of the classes and lectures'. No one, including the writers, knew what that meant (Fox 2004: 541).

The Human Sciences degree seems to have been originally conceived by analogy with PPE and PPP. It was originally administered not by the Anthropology and Geography Board, but by a Committee of the General Board of the Faculties, on which were represented anthropology, sociology, geography, zoology and statistics. Furthermore a measure of academic integration has indeed taken place at ISCA since 1999, with the appointment of the human ecologist and nutritional anthropologist Stanley Ulijaszek, so that Fox's comments must be read in their historical context.

However, I have called my paper 'From Schismogenesis to a New Testament' and I know that it is uppermost in David Parkin's mind that Oxford anthropology should now aim at a new holism, in the spirit of Tylor or Boas – a monistic ideal by contrast with the pluralism of 'human sciences' that seem to have been allowed to coexist amicably without getting in one another's hair. Medical, nutritional and ecological studies are already well in place to be part of this whole, in addition to the themes that I have tried to analyse, plus many other threads of academic continuity that I have not been able to unpick in this short and subjective account. Defying the pluralistic approach to knowledge that is elsewhere so much in vogue, Oxford now has all the ingredients ready to reach even higher excellence than in the past. It remains only for the word to be made flesh.

David Parkin's contribution to the Oxford conference, on the anthropology of crowds (Parkin 2007), is not only highly original but also exemplary in its integrated approach. If successful, his courageous drive towards holism will put to shame those (including myself) who had taken refuge in the idea of social-cultural anthropology as an inescapably marginal discipline whose role is that of an 'underlabourer' to science, as Locke once saw philosophy, with the difference that its labour takes it to the fields rather than remaining like philosophy in the armchair.

Integration of the kind that Parkin aims at will surely require more teamwork than has been usual in anthropology. To borrow from Mary Douglas's typology (Douglas and Wildavsky 1983), the E-P era at Oxford was low-grid, high-group (low regulation, high cohesion), hence tending towards the sectarian. International anthropology today – with increasing thematic and regional specialisation, and easy electronic communication that diminishes the importance of face-to-face interaction – becomes low-grid, low-group: that is to say, a market. Holistic anthropology would call for high-grid, high-group organisation, and hence to hierarchy. In any case, the new initiative should re-establish Oxford as one of the centres of attention for world anthropology.

Acknowledgements

I am grateful for assistance to Brian Street, Hilary Callan, Judith Freedman, Roland Littlewood, Silvia Rodgers and Matt Schaffer, as well as to a number of serving and retired members of the Oxford anthropology community, and in particular to the Editor of this collection. I have also made some use of video interviews with Nicholas Allen and Peter Rivière recorded by Alan Macfarlane and available on his website www.alanmacfarlane.com.

Notes

1. Hilary Callan, pers. comm., quoting an unnamed source.
2. *Le Monde*, 21 December 2004.
3. As noted by Malcolm Chapman (1989: xxiii).
4. It was determined by the university authorities that the Institute was not assigned to the Professor of Social Anthropology.
5. Interview, 7 December 2004.
6. See 'Christianity and British anthropologists: the Oxford Catholic nexus', unpublished paper, c. 1990, deposited in Anthropology Library, British Museum. See also Fardon 1999: 42, 245–6.
7. Other Directors of the Centre have been Soraya Tremayne, Helen Callaway, Cathy Lloyd, Lidia Sciama and (since 2001) Maria Jaschok.
8. Ultimate Frisbee began as a gentle park game on the west coast of the United States but is now played internationally. It is said to mix the best features of sports such as soccer, basketball, American football and netball, and is played without referees.

REFLECTIONS ON OXFORD'S GLOBAL LINKS

Compiled by Wendy James[1]

The growth of anthropology in Oxford, along with virtually all its achievements over the last century, was possible thanks to the comings and goings of staff, students, and ideas in a complex network of two-way links with many countries and institutions across the world. A final panel at the Centenary Conference brought together a sample of people (no statistical reliability claimed!) to reflect on these links between researchers and researched, teachers and taught, in both academic anthropology and its practical applications. The participants' memories certainly testified to a network of connections developing over considerable geographical distances and successive generations.

Three strong and important links were mentioned briefly by Wendy James in introducing the panel. First, the North American, and especially the US link, fostered from the late 1940s by E-P, who spent time himself in Chicago and Stanford, and welcomed many American students and senior guests. The US link has been vital to the thriving of UK anthropology and nowhere more than in Oxford in recent decades. Second, she reminded us of the ancestrally vital connection with the French tradition of Durkheim and Mauss, kept fresh by Louis Dumont's stint at the Institute, and still represented by anthropology's active links with the British Centre for Durkheimian Studies and the Maison Française in Oxford. Third, she mentioned the continuing link with the Nile valley countries. Evans-Pritchard had taught in the University of Cairo in between his main periods of research in the

Sudan in the thirties. After the University of Khartoum was founded in the mid-1950s, several Institute-trained people took posts over the next decade in the department of Social Anthropology and Sociology (including Ian Cunnison, Talal Asad, Lewis Hill, and Wendy herself); and both E-P and Godfrey Lienhardt came out as external examiners. The main contributions to the panel, however, came from people representing a variety of less well-known links, many drawing on personal memories to enrich our celebration of Oxford anthropology's hundredth birthday.

We started by looking back nearly to the 'beginning'. Grażyna Kubica opened the panel on the theme of 'Remembering Maria Czaplicka', an early student of humble background, and of whom she is writing a biography. Czaplicka was a friend of Malinowski, and came from Poland in 1910 to study under Seligman and Westermarck at the LSE. A year later she moved to Oxford, obtaining the Diploma in Anthropology in 1912. In 1914–15, she led the Ienisei Expedition to Siberia, where she became an experienced fieldworker and collected artefacts for the Pitt Rivers Museum. She was the first female lecturer in anthropology at Oxford University (1916–19).

Of Czaplicka's time in Oxford, where she was first accepted by Somerville and later moved to LMH, Grażyna told us:

> In the field of anthropology Czaplicka worked with Marett (she was one of Marett's girls, as they were called). And it was he who discovered that her linguistic and intellectual abilities could be extremely useful ... What was her theoretical profile? She was an adherent to anthropogeography ... She studied for instance the impact of climate, or broadly speaking, environment on religious beliefs ...
>
> She was appointed to the Human Anatomy Department under Professor Arthur Thompson as substitute for a male lecturer (Dudley Buxton) who joined the army. Maria Czaplicka lectured on Ethnology and presented general descriptions of various peoples of Asia and Europe, both in social and physical anthropology (her lecture notes were later used by Beatrice Blackwood ...). This was very typical of female careers during the Great War as they replaced men who fought in the army. But the end was also typical – when the man returned the woman had to resign.
>
> When this happened Czaplicka went to America on a lecture tour, and finally ended up in Bristol as a lecturer in anthropology and planned a centre for anthropological research there. However her difficult financial situation (she had terrible debts) and lack of collegiate support caused her tragic decision to take her life. She wanted to be buried in Oxford. I found her grave at Wolvercote Cemetery (it is in a rather dilapidated state) ...

Czaplicka had joined the RAI and the RGS, among other things, and today there is increased recognition of her contribution with a four-volume set of her works appearing in 1999. Grażyna concluded by mentioning her own welcome in England in 1986 thanks to the Oxford Colleges Hospitality Scheme for Polish Scholars, and she had the opportunity of working under the supervision of Edwin and Shirley Ardener. At this time, coming from a communist country, she found the intellectual climate in Oxford very stimulating; and, we sensed in listening to her, she felt herself in some ways to be recreating the initial experiences of her predecessor.

Africa

Many of the overseas regions that had earlier served as the camping grounds of researchers, but, partly through the upheavals of the Second World War and movements for Independence in India and the colonies, came to be partners in the development of anthropology. Of these, the West African countries and especially Ghana stand out. It was during Meyer Fortes' first presence in Oxford in the early 1940s that this particular link began. Fortes had supervised the work of Kofi Busia, who gained his D.Phil. in 1947. Professor George Hagan, currently Minister of Culture in the Ghanaian government, spoke about the African connections, his own personal experience as a student, and the relevance of anthropology to current questions about the role of culture in development. He noted that the former Eurocentric discourse of anthropology had become a global discourse, and that Oxford had played its part in this by training students from all parts of the world to apply the discipline worldwide, producing 'a handsome crop of that strange hybrid, the African anthro-pologist' – despite the fact that Rattray's works from the colonial period are considered so authentic that many Ghanaians consider him an Asante! He suggested that Ghana holds a special place in the history of Oxford anthropology, 'not because Oxford academics produced classical ethnological works on ethnic groups in Ghana ... but because Oxford produced Kofi Abrefa Busia, a Ghanaian who became the first black African University Professor of Anthropology in the world and, in the early 1970s, the democratically elected Prime Minister of Ghana.' He then asked: 'What did Busia and his successors bring to Anthropology and to Africa?' The first part of his answer was on the 'learning culture' of the Institute.

> Since the end of World War II, Oxford's relatively liberal admissions policy has enabled a stream of African students of varied backgrounds and academic qualifications to come up to Oxford to study at the Institute of Social Anthropology. Back in Africa, they filled an array of service positions ... We lived in the euphoria of the early days of independence.

Over the four years of my studies at the Institute, one encountered
Nigerians, Cameroonians, Egyptians, Sudanese, Ugandans, Kenyans, and a
significant number of Ghanaians. All of us saw ourselves as called to serve
our nation; and Oxford offered a regime of education vaunted as much for
social cultivation as for the cultivation of the mind.

I found my own days as a student of anthropology in Oxford extremely
rewarding and happy. The pressure to produce, week after week, essays
that could stand up to rigorous critical appraisal by masterful and highly
renowned scholars was thankfully relieved by the equally relentless urge to
indulge the soul will the cultural delights that were gratuitously offered on
daily basis both in college and in the Institute and its adjunct pubs by these
same scholars.

Profiling the Institute, George found it 'exotic' – and suggested it
could have passed for the stereotype of a 'primitive' community, one
clan with many lineage groups.

Attachment to the lineage heads was not for academic reasons alone. More
importantly, it was for the general well-being of their broods of students.
The lineage heads feasted and tended us like their own family. The pubs
were the palaver huts where serious academic and thought-provoking
ideas were freely mixed with exciting gossip and serious free counselling.
My first ever visit to the annual Chelsea flower show was at the invitation
of the father of the Ghanaian group, Godfrey; and it was an aesthetic
experience bordering on the transcendental. That was indeed part of the
Oxford experience ...

In the second part of his talk, George Hagan argued for the
relevance of anthropology to development in Africa, specifically
Ghana. It was at Oxford that he read Isaiah Berlin's book, *The Hedgehog
and the Fox* (1953). In the book, using the proverb 'The Hedgehog
knows one big thing; the fox knows many little things', Berlin
distinguished between two kinds of historians: those, on the one hand,
who see a grand design in the movement of world or human events,
such as the Marxists, or even St Augustine, and those who, on the
other hand, see history in episodic terms, emphasising the specificity of
events in terms of time and space, context and particular causes.
George claimed to have left Oxford a fox; but the need to find a way of
showing the relevance of his studies to the contemporary African
situation turned him into more of a hedgehog. Anthropology 'needed
a response to the demands of the nationalist enterprise, just as it had
evolved, at least in part, as a response to the demands of the grand
enterprise of Indirect Rule.' He suggested that of the many studies of
social change that came out in the 1960s and 1970s, David
Brokensha's classic work, originally an Oxford thesis in social

anthropology, *Social Change at Larteh* (1966) provided a model which many African students could use.

In Ghana, at the Institute of African Studies, Hagan and his colleagues began to offer a first-year university course on Culture and Development in the early 1980s, and introduced the topic in a variety of national and international circles, including the Pan-African Anthropological Association and UNESCO. George noted that the following statement can now be found in the Constitution of Ghana: '... The State shall take steps to encourage the integration of appropriate customary values into the fabric of national life through formal and informal education and the conscious introduction of cultural dimensions to relevant aspects of national planning.' This constitutes a watershed in the application of cultural anthropology to development policy; and it also suggests that cultural anthropology might be a requirement in the tool box of the economist or the national planning officer. George Hagan concluded by mentioning a range of current research by Ghanaians which demands the attention of those concerned with practical development. Furthermore, he argued that Africa can 'gain ownership of anthropology' in another way: the 'subjective Afrocentric study of African cultures' can contribute to human enlightenment, as part of a universal discourse concerned with our common humanity and values. African students could best understand their own culture, however, by the study and understanding of other cultures, just as by the study of non-European cultures, Europeans have come to a better understanding of their own.

India

After Africa, the next strong regional link of Oxford anthropology was with India. Professor Ravi Jain, who taught at the Institute from 1966–74, gave an overview of how this had grown since Srinivas was a student of R-B in the 1940s. Ravi Jain entitled his talk 'Oxonian India(s)', and we quote from it at length.

> This is the occasion of celebrating a hundred years of anthropology in Oxford, and I can hardly resist the temptation of narrating, in the first place, the personal trajectory of my involvement in social and cultural anthropology at the Institute in Oxford. I began my career at Oxford in October 1966 as University Lecturer in Indian Sociology and ended in December 1974 from my post then designated as University Lecturer in the Social Anthropology of South Asia and Fellow of Wolfson College. My position at Oxford was a link in an illustrious chain of predecessors: M.N. Srinivas, Louis Dumont and David Pocock and, let me add immediately, my worthy successors – Nick Allen and Marcus Banks ...

I should begin with Srinivas, the apical ancestor of my Oxford academic lineage, who is rightly credited with having initiated the 'field view' rather than the 'book view' of Indian society. Srinivas has himself written in several publications about the impact of Oxford anthropology on his career and ideas, for example, the influence of Africanist dominant clan and dominant lineage ideas to the study of what he conceptualised as 'dominant caste' in multi-caste Indian villages. Indian sociology and social anthropology owe certain of their foundational concepts to Oxford anthropology as transmitted via Srinivas. Let us take for a start Srinivas's *Religion and society among the Coorgs of South India (1952)*. The substantive material for this work came from an earlier Ph.D. thesis completed by Srinivas under Ghurye's supervision at Bombay University. In working this material for an Oxford D. Phil thesis under Radcliffe-Brown's supervision from 1945 onwards, the teaching of the latter, to quote Srinivas, 'greatly modified my approach to the study of human society. At his suggestion, I started applying some of his ideas regarding the inter-relation of religion and society to the data I had already gathered, and this task proved exciting ...'. However Srinivas rejected R-B's alleged anti-historicism and was able to conceptualise the now famous process of 'sanskritisation' and formulate his ideas about the 'spread' of Hinduism. Subsequently Srinivas also disagreed with Evans-Pritchard's stated emphasis on 'design' rather than 'process' in social structural studies and consequently, as his cordial relationship with Gluckman's Manchester approach (grafted from Oxford) was to fructify, even in contemporary India the nuanced distinction between secularism (of political analysts) and 'secularisation' as Srinivas propounded it serves as a guide not only to social scientists but policy-makers as well.

Srinivas also paved the way (as Dumont and Pocock perceived early) for assessing the significance of the purity – impurity opposition in Hindu collective representations. The reference to Dumont and Pocock here is to their jointly authored journal *Contributions to Indian Sociology* (old series) which provided the foundation for what is even today in its New Series avatar or incarnation the leading international journal of socio-cultural anthropology in India.

My own approach to teaching Indian Sociology at Oxford was fairly eclectic. I had not been 'groomed' as Srinivas was by R-B and E-P or Pocock by Dumont into the structuralist mould. But the tutorial system at the Institute fascinated me. There were those tremendously exciting, though differently nuanced textbooks of social anthropology by my Oxford colleagues – Evans- Pritchard, John Beattie, Godfrey Lienhardt and David Pocock. The last named of these being 'foundational' was my bible in the sense that the philosophical rooting/routing of social anthropology in that little book gave me a direction to prepare myself. The fact that Rodney Needham was located physically in the room next to mine in the Institute was a source of counsel and inspiration.

My B.Litt. and D. Phil. students were truly like peers. It was in that spirit that I embarked upon the exciting task of editing, emanating from the decennial meeting of the ASA, the volume called *Text and Context: The*

Social Anthropology of Tradition. I think subliminally I was in the midst of a powerful academic tradition, and I must have wished to synthesise the very rich and mature tradition of Indic society with the intellectual heritage of the European analysts. This I did in my own limited way and with full support of my students and colleagues. Two of my collaborators are here today: Veena Das and Wendy James. Michael Herzfeld at Harvard is another star of that team, and I think his initial training in Indian Sociology and full-blown expertise in studies of Greece later also speaks of the blending or synthesis that I mentioned ...

While making all these observations, I cannot claim to be comprehensive. The latest chapter in Oxonian India(s), for example, is the study of Indian diaspora which engages my full attention now. And, let me remind you, this is nothing new for Oxford. I may not have studied, while in Oxford, say the Indian automobile workers there but I was part of a heritage (having just completed my Ph.D. on Tamil plantation workers in Malaya) which Pocock had initiated with his characteristic sensitivity and erudition. I often tell my pupils in India about the kind of dynamic, contemporary researches that go on inside the mediaeval facade of many Oxford colleges, and if you need an example of this from social anthropology in Oxford just look at the work of Steve Vertovec and his team of Global Networks.

In India we bypassed the currents of nervousness and anxiety, which accompanied the post-modern, 'crisis of representation' in the West. We kept making empirical studies with due regard to reflexivity and the role of agency in human affairs. In relation to Oxford anthropology, though our graduates still study the classics like *The Nuer* and *The Andaman Islanders*, as advanced students they undertake field studies – intensive and fine-grained like they/we have always done – in sociology/anthropology departments in Britain or in India. There is one change though. We do not have the colonial hangover like R-B telling Srinivas at the end of the latter's D. Phil. thesis 'You are not yet ready to teach in India; you still need an apprenticeship of another year of teaching in Oxford.'

Mediterranean

Of the Mediterranean links of Oxford anthropology, Greece was the first, with the appointment of John Peristiany in 1950. The Greek connection has been continuous and important to Oxford ever since, especially with the appointment of John Campbell to a Fellowship at St. Antony's College. Professor Roger Just, originally from Australia, and now of the University of Kent at Canterbury, reflected on the comings and goings. He himself had started with classical Greece, and reminisced on his unlikely arrival at Oxford in 1973, 'as some sort of Australian country boy' on a three-year grant. It was eleven years before he returned, having spent this time between studying in Oxford,

doing fieldwork in Greece, and later working at the British School at
Athens, all of which he enjoyed.

> But I've got a confession to make – I never applied to go to Oxford University,
> actually I didn't want to go to Oxford University. 1973 was the tail end of
> the Vietnam war, which had been a radicalising experience for a whole
> generation of Australian students. Gough Whitlam and the Labour party
> had just been elected after some decades of conservative and obsequiously
> royalist government. And I think we all thought we were radicals, even
> those of us who were probably God's conservatives by nature. And from
> afar, Oxford looked like the bastion of privilege, and the centre of empire.
> Besides which, I had just finished reading *Brideshead Revisited*. I was
> damned if I was walking round the place with a teddy bear stuck under my
> arm.

So Roger applied to the LSE and to SOAS, but neither would take
him because he had no undergraduate degree in anthropology – the
subject was not taught at Melbourne until 1986 (although it was
already established at Sydney and the ANU). In fact the only place in
the UK that would take him to study anthropology, starting with a
postgraduate conversion course, was the Institute, where one started
with the Diploma. So Roger did end up in Oxford, and clearly relieved,
admitted that 'the teddy bear problem never materialised'. He
commented on the surprising variety of people he met.

> ... The fact that [The Institute] didn't have an undergraduate programme,
> made it a curiously open community, a community that allowed people
> such as myself to enter it. It was not only geographically cosmopolitan, but
> as it were intellectually cosmopolitan – in the variety of intellectual
> backgrounds from which people came – I was surrounded by linguists,
> psychologists, historians, zoologists. In a funny way, it was also the centre
> of an erstwhile empire, but I think it was a curiously transformative centre,
> that drew its peripheries into itself, to become part of itself, and which also
> acted as a sort of revolving door through which people passed, met,
> interacted, and were changed.

Roger's Diploma tutor in 1973 was Ravi Jain, even then with
extensive contacts and whose own doctoral studies had been in
Australia. Roger decided on fieldwork in 'modern' Greece for his own
doctorate, and was supervised by Campbell. He left armed with a letter
of introduction from Godfrey Lienhardt to Peristiany, now himself in
Greece after teaching at the Institute largely on the basis of his African
research. Peristiany had been Campbell's own supervisor – another
example of the successive academic generations which underlie
Oxford's global links.

Now all this may be sounding dangerously like an old school tie network, and I suppose to an extent it is that. But in academe the links that are forged, and the influence that is exerted at least relate to ideas, to knowledge, to forms of intellectual enquiry, more than – at least I think so – to actual position. And one way of testing that would be to look at the indirect, rather than the direct, links that Oxford anthropology has made and the influence that it has thereby exerted in Australia and Greece. It would be foolish to say that Australian academe, or Australian anthropology, in any sense has been dominated by actual people who were at Oxford. What's more interesting is the degree to which Oxford anthropology plays its part in Australia despite that.

When in 1990 I finally joined an anthropology programme at the University of Melbourne, my senior colleague was a Texan, though someone who has now spent more of his life in Australia than in Texas. We met, we became good friends, and in the universal manner of academics we sniffed around each other's bookcases and we were both surprised. He had an awful lot of books on Southeast Asia that I didn't, and I had a lot of books on southern Europe that he didn't. But there was a strangely similar core, and in a sense a quite recognisable one: Durkheim, Mauss, Hertz, Hocart, Evans-Pritchard, Fortes, Douglas, Dumont, etc. And then of course the penny dropped. My colleague had never been to Oxford, but he had been taught by Professor Jim Fox! And Professor Jim Fox had been taught by Rodney Needham. And Rodney Needham had taught – amongst very many other people – myself. We shared, across continents, across actually quite different intellectual careers and pasts, a common intellectual capital whose source was the Institute.

I think much the same could be said for anthropology in Greece. I haven't counted up the number of Greek anthropologists who were trained in Oxford; they're not really so very many although a steady stream passed through the hands of John Campbell. I suspect as many went to UCL and LSE. But that lineage of Oxford-trained anthropologists, John Peristiany, his student John Campbell, his students Juliet du Boulay, Renée Hirschon, Michael Herzfeld, plus Renée herself teaching for many years at the University of the Aegean. And through their writings, through *Honour, Family and Patronage*; *Portrait of a Greek Mountain Village*; *Heirs of the Greek Catastrophe*, and Michael's whole library of books has been really established an 'Oxford position' in Greek anthropology, regardless of where people have actually studied. Thank you.

Other regions of the circum-Mediterranean basin also received visiting Oxford anthropologists quite early, with the completion of doctoral theses by Paul Stirling on Turkey and Emrys Peters on Cyrenaica in 1951–2 and by Julian Pitt-Rivers on Spain the following year. In fact Spain has never been far behind in its relation to Oxford, and Professor Carmelo Lisón Tolosana reminded us of the way that the anthropological tradition developed there; his talk was entitled 'The Gardener of Keble Road'. Carmelo spoke of the consolidation of social anthropology as a 'great intellectual creation' of the past century, and an achievement of humanism in bringing together critical reflection on the life of the spirit, through personally-collected ethnography and on the linguistic character of everything human. The Institute in Keble Road struck him as more of a community than an institution.

> I came to anthropology by reading the German diffusionists and the *History of Ethnological Theory* (1937) by Lowie. My anthropological career began at UCL, where I was assigned to a lady tutor, with whom I was not acquainted, called Mary Douglas. She later advised me to transfer to Oxford, where Professor Peristiany taught Anthropology of the Mediterranean.
>
> She wrote to Godfrey Lienhardt and to Professor Evans-Pritchard, and I travelled to Oxford where the Professor had summoned me. On arriving map in hand at 11 Keble Road, I saw a dishevelled man in corduroy trousers pruning a rosebush in the garden. On entering I mentioned to the secretary that the gardener had told me that this was the Institute of Social Anthropology. She interrupted me to explain that it was not the gardener, but Professor Evans-Pritchard. There, though I did not know it, Godfrey Lienhardt was expecting me and invited me to lunch, and it was he who interviewed me, though I did not realise it at the time. Later, he accompanied me to the Professor's office.
>
> I was received kindly: E-P asked about my interest in anthropology and about the existence of the subject in Spain ... Michael Kenny was kind enough to install me in the digs that Dylan Thomas had occupied a few years earlier. With great interest I followed the lectures of John Beattie, who exuded personal experience and serene pragmatism. He taught me the value of detail ... More important still, he suggested that my anthropological education at Oxford also had a moral dimension, which incited me to return to Spain and initiate Anthropology in the University there ...
>
> I never missed the lectures of Evans-Pritchard, whose plural personality impressed me from those first dual encounters in the garden and in his office. In class, he would pass easily from the authoritative gravitas of the professor who has a profound knowledge of the subject he teaches (on entering he would close the door, and nobody was allowed to come in afterwards), to an unexpected and bubbling humour. However, what I really appreciated in his teaching was the dissection and coherence of the ethnographic data, and his comprehension of the sense of the past (which was not appreciated by all his colleagues, but I was a history graduate), as well as his semantic-hermeneutic view of the subject. He recommended me to read Dilthey, with whom I was already acquainted ... He urged me to do fieldwork in Turkey ... I told him I would prefer to go to another region of Spain ... He often spoke of history after one occasion when I mentioned the pirate Drake: 'Pirate?' he asked in surprise. Another time I talked about the Peninsular War, giving a version in which the adventures of Wellington did not appear ... So equipped by Oxford I went to Madrid, where after eight years I was able to intitutionalise the subject in the University, following the basic scheme of the Institute regarding subject-matter and content, but

also taking account of the historical legacy and abundance of sources in all regions of Spain, as well as the riches of a colonial past with hundreds of ethnographic accounts of the Americas dating from the sixteenth and seventeenth centuries. In order to strengthen this anthropological orientation towards Oxford, which at that time I judged to be the most interesting and promising one available, I sent several students to be trained in England ... I started the Library of the Department, with a predominance of English bibliography: works by Evans-Pritchard, Lienhardt, Beattie, Peristiany, Leach and Douglas soon appeared in Spanish ... Those by Evans-Pritchard and Lienhardt are actively recommended today by professors in the Universities of Granada, Corunna, Saragossa, Santander, Madrid, Comillas, San Pablo CEU, among others. With the assistance, at first, of the British Council, I invited up to a couple of dozen British colleagues ...

To summarise: I have had more than sufficient reasons to consider Oxford as my anthropological *alma mater* and the Institute of Social Anthropology as the spring from which I drank ... the most attractive and promising Anthropology, for which I felt, and feel, the greatest affinity and gratitude.

Contacts with the wider Spanish-speaking world spread too, and we were very pleased that Professor Juan Ossio had come all the way from Peru to help celebrate our centenary. It is impossible not to mention too the Oxford – Portugal connection, represented importantly by the work of José Cutileiro and João Pina-Cabral; nor the ventures of Oxford social anthropology into Lowland South America, pioneered by Peter Rivière.

Across the North Sea

The extension of British anthropology in general to the northern European countries, and the Oxford link in particular, developed a little later. Robert Paine studied the Lapps quite early on, with a D.Phil. in 1957. E-P had once offended the Norwegians, by declining to pass Fredrik Barth's Oslo thesis on the Iraqi Kurds, on account of the brevity of the fieldwork and the lack of knowledge of the local language. Barth's subsequent work on the Swat Pathans was awarded a Ph.D. by the LSE, and while Wendy James did spend a period teaching in Bergen (because of the connection established by Barth with the University of Khartoum), it was only after E-P's death that Barth agreed to visit Oxford. There have since been many two-way movements between Oxford and Norway, but the Denmark link has been the most travelled route. Kirsten Hastrup, now Professor of Social Anthropology in the University of Copenhagen, reflected on the story.

In Denmark anthropology was first established as a sub-division of geography in 1945. It was largely based on German culture-history and was located at the National Museum, in keeping with a long tradition of focusing on material culture and linking up with museum ethnography. It grew out of Danish expeditions to central Asia and to the Arctic ... Around 1960, ... the subject was established as an independent discipline in the University of Copenhagen with the establishment of a chair in 1965. The first professor was Johannes Nicolaisen who had worked with the pastoral Tuareg, and who had spent a couple of years at the LSE and had been absolutely smitten by the British way of doing anthropology.

Since then, the royal Oxonian line from Tylor through Radcliffe-Brown to E-P figured prominently on the reading lists. When I started doing anthropology in Copenhagen in the late 1960s, the British functionalist school as it was known to us was supplemented also by French structuralism ... and by slight doses of the Marxism that was to grow in the 1970s.

The first personal link with Oxford, Kirsten explained, was made by Mette Bovin, in the late 1960s a student at the new department of ethnography at Aarhus.

Mette visited Oxford and enthusiastically invited E-P to go back and visit Aarhus. She was completely unimpressed by rank and I think her enthusiasm actually made E-P come. He came back several times, and to this day a guest-room in the very pretty department located outside Aarhus bears his name. There are some wonderful photos of E-P wandering about in the parks of this place ... Since then, what was important for these early links was that to other students in Denmark, Oxford appeared to be a real place with real people and not simply a mythological space somewhere we don't reach.

One year, when E-P was not well enough to go, Wendy went instead. Around this time in the early 1970s, Kirsten saw a notice in *Man* about the launching of *JASO,* and took out a subscription. She also noticed the publication of Edwin Ardener's 1971 edited volume *Social Anthropology and Language,* and 'became aware of a kind of theoretical thinking that went beyond the classification of theories as they were taught in Copenhagen.' Edwin visited Copenhagen in 1973, to help provide an intensive course with other anthropologists and linguists on 'the relation between reality and culture and language and what have you'. Shirley came with him, and Kirsten confessed '... my academic fate was made. In 1974 I came over, and from the first day I marvelled ... I simply marvelled at what I met here.'

First of all, the system of tutorials and supervision, as others have already commented, offered the possibility of concentrated intellectual exchange that allowed one to grow in one's own direction ... And in some sense the vocation to become an anthropologist became personalised through these encounters. Then there were the Friday seminars, and Wednesday coffee mornings, allowing you to talk with people about important matters – or of course less important matters; it all fused into one big experience. There were the fellow students, many of whom are here in this room, and with whom enduring friendships were made ... There was also the Women's Group, established by Shirley, which was a sheltered place where people like myself, who could not always live up to those frightful rhetorical standards of the typical Oxford male, could give a sheltered presentation for the first time ...

In short, and perhaps at a more general level, what I experienced here was an intellectual space that was both generous and supportive of individual ambition. There was, I think, a distinct style of reasoning, starting from an empirical problem and ending at very important theoretical and epistemological issues. I still find that here, to which this very conference has testified, and that I still cherish a lot. There is an openness, a freedom to explore new fields. In my case, it required a shift from studying tribes in India to studying mediaeval Icelandic history for my D.Phil. subject. Everything was possible. And I think perhaps the fact that the departmental identity had become more elusive in the 1970s, which Jonathan Benthall talked about on Friday [see Chapter 8 in this book], provided a window of opportunity for a person like myself to experience anthropology as a total intellectual commitment, even a form of life, where everything mattered and nothing human was beyond interest.

Kirsten, whose own fieldwork has now stretched as far as Greenland, still considered Oxford a point of orientation in her work, and wanted to acknowledge the personal link in education – 'It is through the encounter with real people that one's own academic growth takes shape'. She considered that the Institute taught her that it is very important, in a department, to make room for individual students to develop their own talents, not simply to make them read particular reading lists. 'And I am convinced that the success during the 1990s of my own department in Copenhagen, to which I came in 1990 as the second occupant of the chair of social anthropology in Denmark, is owed not simply to happy conjunctures in Denmark but also to a vision of anthropology which I owe to Oxford.'

Southeast and East Asia

A focus on research methods and dominant questions emerging from thinking at Oxford marked Jim Fox's talk. Under Needham's supervision he found his initial interest in Indonesia; and his commitment to the comparative study of social structure. Since then, his career has blossomed in both predictable and unexpected ways. Jim had come equipped with a professional Powerpoint presentation, but started off by recalling the atmosphere of the Institute in the early sixties, when he first arrived.

> Because for me it was a most exciting time, I had come from Harvard and it was altogether different, the style of teaching, and the strongest memories – as has been often said here in different contexts, were of the pub culture. And at that time in 1962 E-P was in full florescence, and would regularly, I mean regularly, almost every day, go around about twelve o'clock and round up all the students to tell us which pub he would be drinking at and where we had to be in attendance. That was a regular feature throughout the Diploma year. He gave me one piece of advice that I took to heart, it was something he said one time when he was rounding us up, he said 'To get on in your career, never appear to work' – 'but when you're not working, be sure to bother your colleagues'. So he went round and we all went to the pub and usually the afternoons were lost.
>
> But for me ... there were all sorts of exciting ideas at that time ... elementary structures were going to tell us about the fundamentals of society; Lévi-Strauss was still very current and in fact one of the interesting things were the connections that Rodney was building both to Paris and to Leiden University. At the end of Michaelmas Term, with Rodney's urging I went off to Paris, and managed to hear Lévi-Strauss at the Collège de France. He was just beginning the *Mythologiques*, and I sat for three or four lectures totally baffled about how the *cochon sauvage* would relate to elementary structures – and I came back with the word that we were into savage pigs and jaguars and things like that.
>
> But the other element I wanted to emphasise was the idea instilled in all of us that it was a comparative effort. You did ethnography to build towards comparison. And that is really what I took as a major mission, wherever I've gone. In my case, and at Rodney's urging, Eastern Indonesia was the place where all sorts of comparisons could be done. There had been a study, based entirely on mission records and other things in the thirties, a thesis written by a man called van Wouden ... a precursor in many ways to Lévi-Strauss's *Elementary Structures*, and we were all obliged to read that. And that forged the Leiden link. And I remember it was very, very important when the professor from Leiden, Patrick Josselin de Jong, visited Oxford during my Diploma year. It was particularly memorable because at the dinner that Rodney put on for him he announced, before telling me, where I was going to go and do my fieldwork. That's when I first learned that I was going to this island of Roti to do fieldwork. At the time, Clark Cunningham, who went back to the University of Illinois, had done a study on Timor, Rodney had done some limited research on Sumba, and Rodney felt that by sending me to Roti, I would find something in between Sumba and Timor. As it turned out, and it was probably my redemption, I was able to discover in two weeks on the island that we didn't have any of these elementary prescriptive systems, and I could put that aside and do other kinds of ethnography.

Jim then spent a few years in Harvard, where David Maybury-Lewis (another student of Rodney Needham's) was developing a broad comparative study of the peoples of Central Brazil. Moving thence to ANU, Jim has promoted a whole series of comparative studies of the peoples of eastern Indonesia, broadening the concept of what comparison might be by taking into account varieties of the language of social relations within the Austronesian language family as a whole. Doug Lewis (who turned out to be Roger Just's 'Texan colleague') pioneered the series, which continues to this day (as Jim illustrated with detailed maps on the Powerpoint). Jim ended, however, with an ironic story.

> I saw my job as carrying out research in eastern Indonesia, the grand comparative effort in eastern Indonesia, but my first student at the ANU when I arrived in 1975 was a man with the marvellous name of Zamakhsyari Dhofier. He wrote a book on … the Muslim traditional form of schooling, in Java. He went back to Indonesia, and translated it into Indonesian. It happened to sell well, it sold over a hundred thousand copies. And on the first page of his book is this statement, which proclaimed that what he learned was from me, which was hardly true – I guided the thesis, but that produced a flow of students from Indonesia all seeking wisdom at the ANU. In the end I turned out to supervise far, far more theses on Indonesian Islam, on Javanese Islam, than on eastern Indonesia. Now why I say that is because most of the students who came to do preliminary work … would take a reading course that I gave, and the first and foremost reading in that course was Evans-Pritchard's *Sanusi of Cyrenaica*. And I can't tell you how popular that book has become in Indonesian Islamic circles. That is something that has struck, it resonates with a very strong chord, it's greatly appreciated and I believe an Indonesian translation is now under way.

Oxford's eastern links go even further. A China connection was initially made by the appointment of Maurice Freedman to the Chair in 1970. His premature death meant that Chinese studies remained dormant for some time, but with the appointment of Frank Pieke to the new Institute of Chinese Studies in Oxford in the early 1990s, and later that of Elisabeth Hsu to a new post in medical anthropology, the link was revived.

In the case of Japan, the senior scholar Professor Nagashima Nobuhiro pointed to the early dominance of the American anthropological tradition in that country, but indicated how the link with British anthropology, and particularly Oxford, had become strong and significant in recent years. In 1954 two departments were founded, of Social Anthropology at Tokyo Metropolitan University and of Cultural Anthropology at the University of Tokyo. In 1960 some postgraduates at these Universities formed a group of Africanist researchers: Yamaguchi Masao was the leader and Nagashima himself a member. They read extensively in the literature on Africa, much of which came from British social anthropologists, and many of them from Oxford. This period saw a good number of translations of their works into Japanese (including R-B, E-P, Godfrey Lienhardt, Rodney Needham, David Pocock and John Beattie).

The first link with Oxford was through William Newell, one of the last students of R-B, who came to chair the Department of Anthropology and Sociology at International Christian University, Tokyo. Both Yamaguchi and Nagashima himself worked with Newell as his assistant, before leaving for Ibadan and for Oxford respectively. Nagashima was the first Japanese student to come to the Institute in Oxford, and he recalled that in the Japanese *Journal of Ethnology* his own writings were among the few which discussed British social anthropology as a distinctive tradition. By the late 1960s a younger generation of Japanese anthropologists was conducting field research all over the world, its work showing influence from various British scholars.

In 1972 Nagashima Nobuhiro himself founded Social Anthropology at Hitotsubasi University. By the early 1980s the Oxford – Japan link was flourishing; Yoshida Teigo of the University of Tokyo spent half a year in 1983–4 as a Visiting Scholar at the Institute and has remained a useful link ever since. The main counterparts in Oxford who have developed the link are Roger Goodman and Joy Hendry; both have spent long periods in Japan both doing research and teaching. Among Oxford anthropologists who have visited Japan to give lectures or conference papers, we could include Rodney Needham, Robert Barnes, Wendy James, and David Gellner. There has been a steady flow of students in both directions; one Japanese student for example doing fieldwork in the Scottish borders for her D.Phil. (Kaoru Fukuda). It was a particularly interesting development when Akira Okazaki was invited to deliver the first series of Evans-Pritchard commemorative lectures at All Souls College in 1999, on the subject of his own fieldwork in the Sudan on the Ingessana (Gamk) people, originally visited by E-P in 1926.

Kwang-Ok Kim, who had originally arrived in Oxford in 1974, focused on the experience of himself and his wife Okpyo Moon in trying to establish an Oxford tutorial approach in two universities in Korea, where most of their colleagues were trained in American anthropology. Kwang-Ok pointed out that since there were only three Oxford trained anthropologists out of about fifty working in Korea, to talk about any link with Oxford itself seems a little far-fetched. Most had done their post-graduate work in the US. When he returned from Oxford to Korea and started teaching anthropology in the early 1980s, he noticed a number of differences between himself and his colleagues who had been to the States, both in the teaching style and in their approaches to the subject. It was only by noticing those differences, and struggling with them at times, that he began to learn more about some of the distinct features of what may be termed an Oxford tradition, its merits and demerits. Kwang-Ok had been particularly struck by the intensity of the weekly tutorial, backbone of the Institute's teaching. He had been taken aback by the stressful and sometimes humiliating experience of reading an essay aloud to one's tutor, whose role was of course to provide criticism. Against this ordeal, there were more positive experiences.

> There were lectures given at the Institute, but they were not mandatory. In my case, the most awaited moment was the Chinese Seminar on Thursday afternoon organised by the late Professor Maurice Freedman at All Souls. There, I was able to meet and listen to many important scholars working on China, both in England and abroad. After the seminar, we were served splendid afternoon tea with scones and cream, an experience that had impressed me as very British indeed.
>
> I had stayed at Oxford for six years altogether, of which nearly two years were spent among the Taiwanese highland aborigines, Taruko, about whom I wrote my doctoral thesis. I had wanted to study mainland Chinese culture, but Koreans were not yet allowed to enter the Communist China for fieldwork in the late 1970s. It was only since the early 1990s that I started visiting the country regularly for research. Okpyo did her fieldwork in Japan slightly later at the beginning of the 1980s ...
>
> What I want to stress here is that what Oxford offered then was one of the most unsystematic types of teaching, and this put me really at a loss when I started teaching anthropology myself in 1980 back in Korea. ... Compared with me, my colleagues returned from the States at once overwhelmed me with seemingly very sophisticated and systematic lecture syllabuses they brought from the places they studied. They were usually composed of specific weekly topics organised for the whole term and were supplemented with reading lists that appeared to be intimidatingly up-to-date. During my Diploma year, – may be this is just Professor Freedman's own liking – but I was required to read mainly monographs that included such names as Durkheim, Mauss, Hertz, Hubert, Lévy – Bruhl, Dumont, E-P, Mary Douglas, Max Gluckman, Raymond Firth, Edmund Leach, Balandier, and so forth. American-trained colleagues' reading lists, on the other hand, consisted mostly of recent journal articles and a few book chapters. It seemed that an important part of a teacher's work was to guide the students which specific chapters of a book they are supposed to read.
>
> Naturally, I was not very popular among the Korean students at the beginning. ... Moreover, many students seemed to have difficulties adjusting to my emphasis upon lucid but coherent writing rather than extravagant verbal discussion. As time went by and as I was becoming a bit more able to devise my own ways of teaching, I realised and thus became

more appreciative of the fact that the experiences of Oxford tutorials with emphasis upon monograph reading and essay writing had enabled me to develop independent thinking and a more comprehensive grasp of the subject. The painful experiences of exposing oneself to the ruthless criticisms of the supervisor every week, I later realised, effectively trained students to see and accept the exact weaknesses of their arguments.

Kwang-Ok then pondered the apparent gap between the 'heavily metaphysical nature of Oxonian anthropological concerns' and the reality of Korea, liberated from Japanese colonial rule at the end of the Second World War only to be plunged into post-colonial turmoil, and now pursuing rapid development. Korean social scientists were mainly concerned with economic issues such as the elimination of absolute poverty or political ones dealing with the constant instability within the country. However, following the first anthropology department established in 1960, there are now ten. The first focus was on archaeology and folklore studies, but as more anthropologists emerged with American degrees, 'personality studies, economic anthropology and ecological anthropology' dominated the scene. In the 1980s, political anthropology, with notions of dependency, neo-colonial exploitation, and the world system, became key terms that drew most of the students' attention. In this context,

> ... talking about symbolic systems, religious rituals, oral traditions, stateless societies of remote people appeared as something very far away from Korean reality, something morally irresponsible and something that may interest only hobby-seeking dilettantes. I believe that students began to learn that symbols, rituals and religion are issues that are not too removed from reality as they assumed, that they can be important areas for understanding the dynamic relationship between state power, ideological struggles and people's resistance, and that religion and folk beliefs can be investigated as arenas of history and politics. I believe that it was part of my contribution, as the first Korean student who came to Britain and Oxford to study anthropology when everybody went to the United States, to introduce a different kind of tradition – though it is simply my own version of Oxford anthropology – to Korean students. I am pleased to tell that more and more students have become interested in political anthropology, religious anthropology, symbolic anthropology and so forth.

Kwang-Ok concluded by pointing to the fact that very few others followed them from Korea to Oxford, though every year several excellent students leave for the US for higher degrees in anthropology. The problem is the lack of financial support. The network of connections with the UK is a fragile one. In Korea, they tend to write mainly for the Korean audience. However, Kwang-Ok and Okpyo put forward the suggestion that scholars in anthropology across the region could perhaps form a sub-area networking basis in East Asia given the fact that increasingly more students seem to be coming to Britain and Oxford from Taiwan, Singapore, Hong Kong and Japan in recent years.

Varieties of indigenous anthropology: Maori and Muslim perspectives

Anthropology undeniably has its roots partly in the colonial past of Europe, and this was as true of the Oxford variety as of any other, perhaps even more so. But it was the survival and transformation of anthropology in the 'post-imperial' world that ran as a connecting theme through many of our conference contributions. It was the key concern running through the concluding presentations by Ngapare Hopa and Mai Yamani, reflecting on their role as 'indigenous anthropologists' from New Zealand and the Middle East respectively.

Ngapare ['Pare'] Hopa asked in her presentation: 'What happens to research when the insect looks back and the researched become the researchers?' – a question originally posed by the film-maker Merata Mita in 1989. Having toured meetings of the global community of 'indigenous peoples' for some time, and tuning in to their 'voice', and the resulting discourse and critique, she considered what 'the insect' is now doing in New Zealand. She explained it was a serious discourse, the product of many indigenous scholars from around the world, on the effects of colonialism and imperialism on indigenous peoples; on the central role of 'science' and scientific methods, and on the rise of political consciousness that produced the 'modernist resistance struggle' of the post-war years and of the 1960s in particular. She explained how in New Zealand, 'Maori students and scholars of anthropology have come to realise the complicity of "colonial anthropology" and of "colonial history" in the subjugation of our people.' There is emerging in New Zealand a field of research called 'Kaupapa Maori Research' (KMR) that privileges Maori concerns and practices and Maori participation as both researchers and researched, and Pare suggested that this approach means that 'we are no longer the "objects" or insects but the subjects, active in conceptualising our world and concerns.'

> KMR has been defined variously. Here are some examples. KMR is research that is 'culturally 'safe; that involves the mentorship of elders; research that is culturally relevant and appropriate while satisfying the rigours of method, and which is undertaken by a Maori researcher, not a researcher who happens to be Maori ...
>
> Another approach says that KMR 'addresses the prevailing ideologies of cultural superiority' which pervade our social, economic and political

institutions. This is a model framed by the discourses related to the Treaty
of Waitangi and by the development within education of Maori initiatives
which are 'controlled' by Maori. Framing KMR within the Treaty allows
space for the involvement of non-Maori in support of Maori research.

Pare summarised the main feature of this approach by saying that
KMR is 'related to "being" Maori'; it is connected to Maori philosophy
and principles, and takes for granted the validity of Maori language
and culture; it is concerned with the 'struggle' for autonomy over the
cultural well-being of the people. As part of the wider context, KMR
has been located within the wider project of *Kaupapa Maori*, of which
the basic elements/principles include: respect for extended family
structure, cultural aspirations, and the aim of a collective vision.

A final point Pare put forward about KMR is its relation to critical
theory, in particular to the notions of critique, resistance, struggle and
emancipation. It has led to situating Maori research within the anti-
positivist debate raised by critical theory and to the declaration that

> ... intrinsic to Kaupapa Maori theory is an analysis of existing power
> structures and societal inequalities. Kaupapa Maori theory therefore aligns
> with critical theory in the act of exposing underlying assumptions that
> serve to conceal the power relations that exist within society and the ways
> in which dominant groups construct concepts of 'common sense' and
> 'facts' to provide *ad hoc* justification for the maintenance of inequalities
> and the continued oppression of Maori people.

We will appreciate, she concluded, how much this approach
appeals to the insect!

The theme of 'anthropology at home' rounded off the panel
presentations, with a poignant description by Mai Yamani of her work
as a Saudi Arabian anthropologist able, at first, to study only the elite
of her own country; but then to move into international research and
inevitably encounter the political turmoils of the wider world.

> Anthropology has changed; Islam also has changed. We have witnessed
> tremendous developments in anthropology. Anthropology is no longer
> simply the western study of the other, but has become a part of how all
> cultures understand themselves ...
>
> I would like to offer my experiences, as an Arab Muslim, on an
> anthropological mission to reveal the vibrant diversity in the Arab world –
> a diversity that is under constant threat of suppression. Perhaps it would be
> elucidating if I offer myself as a case study. I came to Oxford in 1979 as a
> postgraduate student. I saw myself, I saw my identity, as a Muslim, Saudi,
> Iraqi woman – Iraqi mother, Saudi father of Yemeni origin born in Cairo;
> and I grew up in Mecca, in the Hejaz – so that is Arab diversity. My life as a
> student of anthropology at the Institute in Banbury Road became divided

between the library and the Horse and Jockey pub. Although I originate from Mecca, the melting pot of the Muslim world, the ethnic diversity that I encountered in Oxford was of a deeper awareness and understanding. At Oxford we celebrate our differences. As I sat with Peter and Godfrey Lienhardt, my supervisors later – first Peter and then Godfrey, who in his own words inherited me – discussing the tribes of the southern Sudan, the tribes of the Arabian peninsula, oral traditions, cargo cults, leopard skin chiefs, over traditional English refreshments, Islam was undergoing revolution, revival, and war. In Iran, Khomeini leading a Shi'a revolution was threatening the hegemony of the Wahhabi Sunni neighbours in Saudi Arabia, and questioning their custodianships of Mecca and Medina. The Saudi rulers retaliated by launching an aggressive, militant Sunni Wahhabi Islam. Saudi Arabia intensified its religious dogma. School curricula became more Islamic; radio and television programmes carried more Islamic messages, and members of the official committee for the ordering of the good and the forbidding of the evil, the ... religious police, were unleashed on citizens with renewed vigour. They patrolled in their jeeps the streets of the kingdom, searching for sin. Sins were not hard to find ...

I left Oxford in 1981 after completing the Diploma in Anthropology, to become a lecturer in social anthropology at King Abdul Aziz University in Jeddah. I was the first woman, Saudi woman lecturer, to lecture to women, according to the strict segregation of the University in Jeddah. I tried with all the growing interest and enthusiasm to introduce ideas of respect and of cultural diversity. I got in all the books I found translated into Arabic from Oxford, such as Evans-Pritchard's *Social Anthropology*. Although so many of my female students responded to these exciting and exotic themes and concepts, official censorship was stifling. And the compulsion there became heavier and heavier, both physically and emotionally. So I came back to Oxford in search of academic freedom and the opportunity to understand my own background.

Mai's subsequent D. Phil. thesis on the Hejazi identity, she explained, was both an academic pursuit and a personal quest. She then continued research in Saudi Arabia, developing a deepening connection with the minorities. The Shi'a in the eastern provinces, who are discriminated against; the Hejazis, who also are marginalised; the youth in that country. Her books, published in London, breached the official lines of censorship, and were banned; as was she. As a research fellow of the Royal Institute of International Affairs in London, she is nevertheless continuing her research, at present on Muslim communities in Britain, and pursuing her 'mission of defending cultural dignity and freedom of expression'. Mai is often recruited, too, as a commentator on British, and worldwide, radio and television news programmes.

The dilemmas of Mai Yamani's position reminded her listeners of the way that Maria Czaplicka found a voice for her own work on coming to Oxford in 1910 and finding the breadth of possibilities offered here; not to mention her biographer, Grażyna Kubica's experience of visiting from communist Poland in the 1980s. Roger Just, who presented himself as something of a country boy from the Australian outback, had expected to find Oxford in the 1970s a bastion of colonial empire and its ageing defenders, and was surprised to find that the empire had collapsed on itself and Oxford was full of lively and ambitious people from all over the world. George Hagan testified to the importance of the warmth and personal support that he and other students from Africa had found in Oxford, at the Institute in particular. Testimonies of this kind reminded us of the most important reason why we undertook the Centenary celebrations in the first place-not only as an academic event but as a reunion and a chance to take the older connections forward. The testimonies included here from the living 'diaspora' also complement the main chapters of this volume which deal with the internal 'past' history of the discipline, as a teaching and research tradition, managing, if only just, to put down roots and survive institutionally over the span of a century in Oxford itself.

Note

1.   In compiling this account I have drawn on written notes prepared by several speakers at the final panel of the Centenary Conference, and also a video recording of most of the event made by Alan Macfarlane.

BIBLIOGRAPHY

Adler, J., R. Fardon and C. Tully. 2003. *From Prague Poet to Oxford Anthropologist: Frank Baermann Steiner Celebrated: Essays and Translations*. London: Institute of Germanic Studies, University of London.

Allen, N. 2003. From mountains to mythologies. *Ethnos* 68: 271–84.

Al-Shahi, A. 1999. Evans-Pritchard, Anthropology, and Catholicism at Oxford: Godfrey Lienhardt's view. *JASO* 30: 67–72.

―――― and F.C.T. Moore (trans and eds). 1978. *Wisdom from the Nile: a Collection of Folk-stories from Northern and Central Sudan*. Oxford: Clarendon Press.

Andrzejewski, B.W. and I.M. Lewis. 1964. *Somali Poetry: an Introduction*. Oxford: Clarendon Press.

Anon. 1867. Oxford Anthropological Society. *Anthropological Review* 5: 372–3.

Ardener, E. 1971. The new anthropology and its critics. *Man* (n.s.) 6: 449–67.

―――― 1980. Ten years of JASO. *JASO* 11: 124–31.

―――― 1989 [1971]. Social anthropology and language. In *The Voice of Prophecy and Other Essays*, M. Chapman (ed.). Oxford: Blackwell.

―――― 1989 [1973]. Behaviour – a social anthropological criticism. In *The Voice of Prophecy and Other Essays*, M. Chapman (ed.). Oxford: Blackwell.

―――― 1989 [1987]. 'Remote areas'– some theoretical considerations. In *The Voice of Prophecy and Other Essays*, M. Chapman (ed.). Oxford: Blackwell.

Ardener, E. (ed.). 1971. *Social Anthropology and Language*. ASA Monograph 10. London: Tavistock.

Ardener, S. 1985. The social anthropology of women and feminist anthropology. *Anthropology Today* 1/5: 24–6.

―――― 2005. Ardener's 'muted groups': the genesis of an idea and its praxis. Unpublished paper.

―――― (ed.). 1975. *Perceiving Women*. London: Dent.

―――― (ed.). 1981. *Defining Females*. London: Croom Helm.

Asad, T. (ed.) 1973. *Anthropology and the Colonial Encounter*. London: Ithaca Press.

Babalola, S.A. 1966. *The Content and Form of Yoruba ijala*. Oxford: Clarendon Press.

Bailey, J.D. n.d. The University Years, 1914–45. Typescript. All Souls College.

Barnes, J.A. 1987. Edward Evan Evans-Pritchard. *Proceedings of the British Academy* 73: 447–89.

Barnes, R.H. 1974. *Kédang: A Study of the Collective Thought of an Eastern Indonesian People*. Oxford: Clarendon Press.

Beattie, J.H.M. 1964. *Other Cultures: Aims, Methods and Achievements of Social Anthropology*. London: Cohen & West.

Beidelman, T.O. 1974. *A Bibliography of the Writings of E.E. Evans-Pritchard*. Compiled by E.E. Evans-Pritchard, amended and corrected by T.O. Beidelman. London: Tavistock.

Benthall, J. (ed.) 2002. *The Best of Anthropology Today*. London: Routledge.

Berlin, I. 1953. *The Hedgehog and the Fox: An Essay on Tolstoy's View of History*. London: Weidenfeld & Nicolson.

——— 1980. 'Richard Pares'. In *Personal Impressions*, H. Hardy (ed.). Oxford: Oxford University Press.

——— 2001. The state of psychology in 1936. *History and Philosophy of Psychology* 3: 76–83.

Bohannan, P. 1963. *Introduction to Social Anthropology*. New York and London: Holt, Rinehart & Winston.

Bourdillon, M.F.C. and M. Fortes (eds). 1980. *Sacrifice*. London: Academic Press for RAI.

Boyce, A.J. and C.G.N. Mascie-Taylor (eds). 1996. *Molecular Biology and Human Diversity*. Cambridge: Cambridge University Press.

Brice, W. (ed.). 1953. *Anthropology at Oxford: The Proceedings of the Five – hundredth Meeting of the Oxford University Anthropological Society, held on February 25th, 1953, under the Chairmanship of Sir Alan Pim*. Oxford: Holywell Press.

Brokensha, D. 1966. *Social Change at Larteh, Ghana*. Oxford: Clarendon Press.

Brown, A., J. Coote and C. Gosden. 2000. Tylor's tongue: Material culture, evidence and social networks. *JASO* 31: 257–76.

Burton, J.W. 1992. *An Introduction to Evans-Pritchard*. Studia Instituti Anthropos 45. Fribourg: University Press.

Busia, K.A. 1951. *The Position of the Chief in the Modern Political System of Ashanti*. London: Oxford University Press for the International African Institute.

Buxton, J. 1963. *Chiefs and Strangers: A Study of Political Assimilation among the Mandari*. Oxford: Clarendon Press.

Buxton, L.H.D. 1924. *The Eastern Road*. London: Kegan Paul.

——— (ed.). 1936. *Custom is King: Essays presented to R.R. Marett*. London: Hutchinson.

Callan, H. 1970. *Ethology and Society: Towards an Anthropological View*. Oxford: Clarendon Press.

Campbell, J.K. 1964. *Honour, Family and Patronage. A Study of Institutions and Values in a Greek Mountain Community*. Oxford: Clarendon Press.

Chapman, M. (ed.). 1989. *The Voice of Prophecy and Other Essays*. Oxford: Blackwell.

Chapman, W.R. 1982. Ethnology in the museum: A.H.L.F. Pitt Rivers (1827–1900) and the institutional foundations of British anthropology. D. Phil. Thesis, University of Oxford.

—— 1984. Pitt Rivers and his collection. In *The General's Gift: A Celebration of the Pitt Rivers Museum Centenary 1884–1984*, B.A.L. Cranstone and S. Seidenberg (eds). *JASO* Occasional Paper 3.

Collingwood, R.G. 1939. *An Autobiography*. Oxford: Clarendon Press.

—— 1945. *The Idea of Nature*. Oxford: Clarendon Press.

—— 1946. *The Idea of History*. Oxford: Clarendon Press (Revised edn 1993).

—— 2005. *The Philosophy of Enchantment: Studies in Folktale, Cultural Criticism, and Anthropology*, D. Boucher, W. James and P. Smallwood (eds). Oxford: Clarendon Press.

Colville, J. 2004. *The Fringes of Power. Downing St. Diaries, 1939–55* (Revised edition). London: Weidenfeld & Nicolson.

Coote, J. 1981. Selections from the minutes of the Oxford University Anthropological Society. Members past and present: 1. Professor Meyer Fortes. *JASO* 12: 114–15.

—— 1982. Selections from the minutes of the Oxford University Anthropological Society. Members past and present: 2. Dr Audrey Richards. *JASO* 13: 110–11.

—— 1985. Selections from the minutes of the Oxford University Anthropological Society. Members past and present: 3. Professor Max Gluckman. *JASO* 16: 143–4.

—— 1991. J.H.M. Beattie and the OUAS. *JASO* 22: 73.

—— and A. Shelton (eds). 1992. *Anthropology, Art and Aesthetics*. Oxford: Clarendon Press.

Cranstone, B.A.L. and S. Seidenberg (eds). 1984. *The General's Gift. A Celebration of the Pitt Rivers Museum Centenary 1884–1984. JASO* Occasional Papers 3.

Curthoys, M.C. 1997. The examination system. In *The History of the University of Oxford. (Volume VI) Nineteenth Century Oxford, Part I*, M. G. Brock and M. C. Curthoys (eds). Oxford: Clarendon Press.

Cutileiro, J.P.1971. *A Portuguese Rural Society*. Oxford: Clarendon Press.

Czaplicka, M.A. 1914. *Aboriginal Siberia. A Study in Social Anthropology*. Oxford: Clarendon Press.

—— 1916. *My Siberian Year*. London: Mills & Boon.

Davin, D. 1972. *Brides of Price*. London: Robert Hale.

Davis, J. 1992. The anthropology of suffering. *Journal of Refugee Studies* 5:149–61.

Deng, F.M. 1973. *The Dinka and their Songs*. Oxford: Clarendon Press.

Descola, P. 1994. *In the Society of Nature: a Native Ecology of Amazonia*. Cambridge: Cambridge University Press.

Douglas, M. 1966. *Purity and Danger*. London: Routledge & Kegan Paul.

—— 1980. *Evans-Pritchard*. London: Fontana.

—— and A. Wildavsky. 1983. *Risk and Culture*. Berkeley: University of California Press.

Dresch, P. and W. James. 2000. Introduction: Fieldwork and the passage of time. In *Anthropologists in a Wider World: Essays on Field Research*, P. Dresch, W. James and D. Parkin (eds). New York and Oxford: Berghahn Books.

Du Boulay, J. 1974. *Portrait of a Greek Mountain Village*. Oxford: Clarendon Press.

Dunbabin, J.P.D. 1994. Finance since 1914. In *The History of the University of Oxford. (Volume VIII). The Twentieth Century*, B. Harrison (ed.). Oxford: Clarendon Press

Edwards, E. 2001. *Raw Histories: Photographs, Anthropology and Museums*. London: Routledge.

―――― (ed.). 1992. *Anthropology and Photography 1860–1920*. London and Newhaven: Yale University Press.

Evans-Pritchard, E.E. 1937a. *Witchcraft, Oracles and Magic among the Azande*. Oxford: Clarendon Press.

―――― 1937b. Anthropology and administration. *Oxford Summer School on Colonial Administration*. Oxford: Oxford University Press.

―――― 1938. Some administrative problems in the Southern Sudan. *Oxford Summer School on Colonial Administration*. Oxford: Oxford University Press.

―――― 1940. *The Nuer: a Description of the Modes of Livelihood and Political Institutions of a Nilotic People*. Oxford: Clarendon Press.

―――― 1949. *The Sanusi of Cyrenaica*. Oxford: Clarendon Press

―――― 1950. Social anthropology: Past and present [The Marett Lecture], repr.in *Essays in Social Anthropology*, 1962. London: Faber and Faber.

―――― 1951a. *Social Anthropology*. London: Cohen & West.

―――― 1951b. The Institute of Social Anthropology. *Oxford Magazine*, 26 April.

―――― 1956. *Nuer Religion*. Oxford: Clarendon Press.

―――― 1959. The teaching of social anthropology at Oxford. *Man* 64: 121–4.

―――― 1973. Fifty years of British anthropology. *TLS*, 6 July: 763–4.

―――― 1981. *A History of Anthropological Thought*. A. Singer (ed.), with intro. by E. Gellner. London: Faber and Faber.

―――― (ed.). 1954. *The Institutions of Primitive Society*. London: Cohen & West.

Fardon, R. 1999. *Mary Douglas: An Intellectual Biography*. London: Routledge.

Finnegan, R.H. 1967. *Limba Stories and Story-telling*. Oxford: Clarendon Press.

―――― 1970. *Oral Literature in Africa*. Oxford: Clarendon Press.

Firth, R. 1981. Spiritual aroma: religion and politics. AAA Distinguished Lecture. *American Anthropologist* 83: 582.

Fortes, M. and E.E. Evans-Pritchard (eds). 1940. *African Political Systems*. London: Oxford University Press.

Fox, Robert 1997. The University Museum and Oxford science, 1850–1880. In *The History of the University of Oxford. (Volume VI) Nineteenth Century Oxford, Part I*, M.G. Brock and M.C. Curthoys (eds). Oxford: Clarendon Press.

Fox, Robin. 2004. *Participant Observer: Memoir of a Transatlantic Life*. New Brunswick: Transaction.

Frazer, J.G. 1913. The scope of social anthropology. In *Pysche's task: A Discourse concerning the Influence of Superstition on the Growth of Institutions*. London: Macmillan.

Galtung, J. 1969. Violence, peace and peace research. *Journal of Peace Research* 6: 167–91.

Geertz, C. 1988. *Works and Lives: The Anthropologist as Author*. Stanford, CA: Stanford University Press.

Gellner, E. 1981. Introduction to Evans-Pritchard. In *A History of Anthropological Thought*, A. Singer (ed.). London: Faber and Faber.

Gilsenan, M. 1973. *Saint and Sufi in Modern Egypt: An Essay in the Sociology of Religion*. Oxford: Clarendon Press.

Goody, J. 1972. *The Myth of the Bagre*. Oxford: Clarendon Press.

———— 1995. *The Expansive Moment: Anthropology in Britain and Africa, 1918–1970*. Cambridge: Cambridge University Press.

Gosden, C., F. Larson and A. Petch. 2007. *What is a Museum? Exploring the Collections of the Pitt Rivers Museum 1884–1945*. Oxford: Oxford University Press.

Green, S.J.D. 2005. Scholars and imperialists. Typescript. All Souls College.

Hailey, L. 1938. *An African Survey: a Study of Problems arising in Africa South of the Sahara*. Oxford: Oxford University Press.

Harrell-Bond, B.E. 1986. *Imposing Aid: Emergency Assistance to Refugees*. Oxford: Oxford University Press.

———— and E. Voutira. 1992. Anthropology and the study of refugees. *Anthropology Today* 8/4: 6–10.

Harrison, B. 1994. *The History of the University of Oxford. (Volume VIII). The Twentieth Century*. Oxford: Clarendon Press.

Harrison, G.A. 1995. *The Human Biology of the English Village*. Oxford: Oxford University Press.

———— and A.J. Boyce (eds). 1972. *The Structure of Human Populations*. Oxford: Oxford University Press.

———— and J.B. Gibson (eds). 1976. *Man in Urban Environments*. Oxford: Oxford University Press.

———— J.S. Weiner, J.M. Tanner and N.A. Barnicot. 1964. *Human Biology*. Oxford: Oxford University Press.

Hastrup, K. 1993. Hunger and the hardness of facts. *Man* (n.s.) 28: 727–39.

Henson, H. 1974. *British Social Anthropologists and Language: A History of Separate Development*. Oxford: Clarendon Press.

Herle, A. 1998. The life-histories of objects: collections of the Cambridge anthropological expedition to the Torres Strait. In *Cambridge and the Torres Strait: Centenary Essays on the 1898 Anthropological Expedition*, A. Herle and S. Rouse (eds). Cambridge: Cambridge University Press.

Hirschon, R. 1989. *Heirs of the Greek Catastrophe. The Social Life of Asia Minor Refugees in Piraeus*. Oxford: Clarendon Press.

Hitchcock, M. 1980a. Selections from the minutes of the Oxford University Anthropological Society. 1909: The magic lantern and the 'Messrs.' of Oriel and Lincoln. *JASO* 11: 96–8.

———— 1980b. Selections from the minutes of the Oxford University Anthropological Society. 1919–20: A diet of skulls, women and cannibals. *JASO* 11: 170–1.

———— 1981. Selections from the minutes of the Oxford University Anthropological Society. *JASO* 12: 14–15.

Hubert, H. 1999 [1905]. *Essay on Time: A Brief Study of the Representation of Time in Religion and Magic*, R. Parkin and J. Redding (trans). Oxford: Durkheim Press/Berghahn Books.

Ingold, T. 2000. *The Perception of the Environment: Essays on Livelihood, Dwelling and Skill*. London: Routledge.

Ishiguro, K. 1986. *An Artist of the Floating World*. London: Faber and Faber.

Jain, R. (ed.). 1977. *Text and Context. The Social Anthropology of Tradition*. ASA Essays in Social Anthropology 2. Philadelphia: Institute for the Study of Human Issues.

James, W. 1988. *Listening Ebony: Moral Knowledge, Religion and Power among the Uduk of Sudan*. Oxford: Clarendon Press.

—— 2003. *The Ceremonial Animal*. Oxford: Oxford University Press.

—— 2005. A fieldworker's philosopher: Perspectives from anthropology. In *The Philosophy of Enchantment*, D. Boucher, W. James, and P. Smallwood (eds). Oxford: Clarendon Press.

Johnson, D.H. 1982. Evans-Pritchard, the Nuer and the Sudan Political Service. *African Affairs* 81: 231–46.

—— 2007. Political intelligence, colonial ethnography and analytical anthropology in the Sudan. In *Ordering of Africa: Anthropology, European Imperialism and the Politics of Knowledge*, H. Tilley and R. Gordon (eds). Manchester: Manchester University Press.

Just, R. 1989. *Women in Athenian Law and Life*. London: Routledge.

Keuren, D.K. van. 1991. From natural history to social science. In *The Estate of Social Knowledge*, J. Brown and D.K. van Keuren (eds). Baltimore and London: Johns Hopkins University Press.

Krapf-Askari, E. [later Gillies] 1969. *Yoruba Towns and Cities: An Enquiry into the Nature of Urban Social Phenomena*. Oxford: Clarendon Press.

Kuklick, H. 1991. *The Savage Within. The Social History of British Anthropology, 1885–1945*. Cambridge: Cambridge University Press.

Kunene, D.P. 1971. *Heroic Poetry of the Basotho*. Oxford: Clarendon Press.

Kuper, A. 1973. *Anthropologists and Anthropology: The British School 1922–1972*. London: Allen Lane.

Latour, B. 1993. *We have Never been Modern*. London: Harvester Wheatsheaf.

—— 2004. *Politics of Nature: How to bring the Sciences into Democracy*. Cambridge, Mass: Harvard University Press.

Lavin, D. 1995. *From Empire to International Commonwealth: a Biography of Lionel Curtis*. Oxford: Clarendon Press.

Leach, E. 1984. Glimpses of the unmentionable in the history of British social anthropology. *Annual Review of Anthropology* 13: 1–23.

Le Gros Clark, W.E. 1934. *Early Forerunners of Man*. London: Bailliere, Tindall and Cox.

—— 1955. *The Fossil Evidence for Human Evolution*. Chicago: Chicago University Press.

Lévi–Strauss, C. 1989. Letter. *Anthropology Today* 5/2: 23

Lienhardt, P. (ed. and trans.). 1968. *Hasani Bin Ismail. The Medicine Man: Swifa ya Nguvumali*. Oxford: Clarendon Press.

Lienhardt, R.G. 1961. *Divinity and Experience: The Religion of the Dinka*. Oxford: Clarendon Press.

—— 1964. *Social Anthropology*. London: OUP for the Home University Library.

—— 1974. E-P: A personal view. *Man* (n.s.) 9: 299–304.

Loizos, P. 1981. *The Heart grown Bitter: A Chronicle of Cypriot War Refugees.* Cambridge: Cambridge University Press.

Lowie, R.H. 1937. *The History of Ethnological Theory.* London: Harrap.

Lubbock, J. 1870. *The Origin of Civilisation and the Primitive Condition of Man.* London: Longmans, Green, and Co.

Marett, R.R. 1909. *The Threshold of Religion.* London: Methuen.

——— 1912. *Anthropology.* London: Williams and Norgate.

——— 1920. *Psychology and Folklore.* London: Methuen.

——— 1929. Anthropology. *Encyclopaedia Britannica* (14th edition) 2: 41–6.

——— 1932. *Faith, Hope and Charity in Primitive Religion.* Oxford: Clarendon Press.

——— 1935. *Head, Heart, and Hands in Human Evolution.* London: Hutchinson.

——— 1936. *Modern Sociologists – Tylor.* New York: John Wiley and Sons, Inc.

——— 1941. *A Jerseyman at Oxford.* Oxford: Oxford University Press.

——— 1997. *The Threshold of Religion.* London: Routledge/Thoemmes Press.

——— (ed.). 1908. *Anthropology and the Classics.* Oxford: Clarendon Press.

Mauss, M. 1954 [1925]. *The Gift: Forms and Functions of Exchange in Archaic Societies,* I. Cunnison (tran.). London: Cohen & West.

——— 1994. Four letters to Radcliffe-Brown from Durkheim and Mauss. JASO 25: 169–78.

——— 2003 [1909]. *On Prayer.* W.S.F. Pickering (ed.), with concluding remarks by H. Morphy, S. Leslie (tran.). Oxford: Durkheim Press/Berghahn Books.

——— 2005. *Techniques, Technology and Civilisation.* N. Schlanger (ed. & tran.). Oxford: Durkheim Press/Berghahn Books.

Mbiti, J.S. 1966. *Akamba Stories.* Oxford: Clarendon Press.

Mills, D. 2003a. Quantifying the discipline. Some anthropology statistics from the UK. *Anthropology Today* 19/3: 19–22.

——— 2003b. Professionalizing or popularizing anthropology? A brief history of anthropology's scholarly associations in the UK. *Anthropology Today* 19/5: 8–13.

Milton, K. 1996. *Environmentalism and Cultural Theory.* London: Routledge.

Minority Rights Group. 1980. *The Refugee Dilemma: International Recognition and Acceptance.* Report no. 43, F. D'Souza (ed.). London: Minority Rights Group.

Morphy, H. and M. Banks. 1997. *Rethinking Visual Anthropology.* New Haven and London: Yale University Press.

Morrell, J.B. 1994. The non–medical sciences, 1914–1939. In *The History of the University of Oxford. (Volume VIII). The Twentieth Century,* B. Harrison (ed.). Oxford: Clarendon Press.

Myres, J.L. 1953. Memories of Sir Edward Tylor. In *Anthropology at Oxford: The Proceedings of the five–hundredth Meeting of the Oxford University Anthropological Society, held on February 25th, 1953, under the chairmanship of Sir Alan Pim,* W. Brice (ed.). Oxford: Holywell Press.

Needham, R. 1962. *Structure and Sentiment.* Chicago: Chicago University Press.

——— 1964. Reorganization of anthropology at Oxford. *Man* 64: 153–4.

—— 1975. Polythetic classification: convergence and consequences. *Man* (n.s.) 10: 349–69.

—— 1979. Video: Anthropologist Rodney Needham interviewed by James J. Fox on the ANU campus, Canberra, September 1979. Canberra: Ethnographic Film Unit, Research School of Pacific and Asian Studies, Australian National University.

—— 1985. *Exemplars*. Berkeley: University of California Press.

Parkin, D. 2007. The social in the visceral: Development and transformations of the crowd. In *Holistic Anthropology: Emergence and Convergence*, D. Parkin and S. Ulijaszek (eds). New York and Oxford: Berghahn Books.

—— and S. Ulijaszek (eds). 2007. *Holistic Anthropology: Emergence and Convergence*. New York and Oxford: Berghahn Books.

Parkin, R.J. 1983. Lévi–Strauss and the Austroasiatics: 'Elementary structures' under the microscope. JASO 14: 79–86.

Pocock, D. 1961. *Social Anthropology*. London: Sheed & Ward.

Radcliffe-Brown, A.R. 1952. *Structure and Function in Primitive Society*. London: Cohen & West.

—— 1985. Footnotes for the history of anthropology. *History of Anthropology Newsletter* 12/2: 3–11.

—— and D. Forde (eds). 1950. *African Systems of Kinship and Marriage*. London: Oxford University Press.

Reiter, R. 1975. *Toward an Anthropology of Women*. New York: Monthly Review Press.

Reynolds, V. 1976. *The Biology of Human Action*. Reading: Freeman.

—— 2005. *The Chimpanzees of the Budongo Forest*. Oxford: Oxford University Press.

—— and R.E.S. Tanner. 1983. *The Biology of Religion*. London: Longmans.

Rivière, P. 1974. The couvade: a problem reborn. *Man* (n.s.) 9: 423–35.

—— 1985. Unscrambling parenthood: the Warnock Report. *Anthropology Today* 1/4: 2–7.

Rodgers, S. 1984. Feminine power at sea. *Royal Anthropological Institute News* 64: 2–4.

Rosaldo, M.Z. and L. Lamphere. 1974. *Woman, Culture, and Society*. Stanford: Stanford University Press.

Rowse, A.L. 1962. Richard Pares. *Proceedings of the British Academy* 48: 345–56.

Schaffer, M. 2003. *Djinns, Stars and Warriors: Mandinka Legends from Pakao, Senegal*. Leiden: Brill.

—— 2005. Bound to Africa: the Mandinka legacy in the New World. *History in Africa* 32: 321–69.

—— and C. Cooper. 1987. *Mandinko, The Ethnography of a West African Holy Land*. Prospect Heights, IL.: Waveland Press. (First published in 1980 by Holt, Rinehart and Winston).

Simpson, R. 1983. *How the PhD came to Britain. A Century of Struggle for Postgraduate Education*. Society for Research into Higher Education, University of Surrey: Guildford.

Srinivas, M.N. 1952. *Religion and Society among the Coorgs of South India*. Oxford: Clarendon Press.

Steiner, F.B. 1999a. *Selected Writings*. J. Adler and R. Fardon (eds). New York and Oxford: Berghahn Books.

———— 1999b. *Taboo*. J. Adler and R. Fardon (eds). Oxford: Berghahn Books.

Stocking, G.W. 1995. *After Tylor. British Social Anthropology 1888–1951*. Madison: University of Wisconsin Press.

Temple, R.C. 1913. Report of a discussion on 'The practical application of anthropological teaching in universities'. *Man* 13: 185–92.

———— 1914. Anthropological teaching in the universities. *Man* 14: 57–72.

———— 1921. 'Tout savoir, tout pardonner'. An appeal for an Imperial School of Applied Anthropology. *Man* 21: 150–5, 173–5.

Trautmann, T.R. 1992. The revolution in ethnological time. *Man* (n.s.) 27: 379–97.

Tylor, E.B. 1871. *Primitive Culture: Researches into the Development of Mythology, Philosophy, Religion, Language, Art, and Custom* (2 vols). London: John Murray.

———— 1889. On a method of investigating the development of institutions: Applied to laws of marriage and descent. *Journal of the Anthropological Institute* 18: 245–72.

Ulijaszek, S.J. and R.A. Huss-Ashmore (eds). 1997. *Human Adaptability Past, Present and Future*. Oxford: Oxford University Press.

Wallis, W.D. 1957. Anthropology in England early in the present century. *American Anthropologist* 59: 781–90.

Webber, J. 1983. *Register of Theses in Social Anthropology accepted for Higher Degrees in British Universities 1975–80*. London: Royal Anthropological Institute.

Weiner, J.S. 1955. *The Piltdown Forgery*. London: Oxford University Press.

Willis, R. 1978. *There was a Certain Man: Spoken Art of the Fipa*. Oxford: Clarendon Press.

Wilson, B.R. (ed.). 1970. *Rationality*. Oxford: Blackwell.

Winch, P. 1971. *The Idea of a Social Science and its Relation to Philosophy*. London: Routledge and Kegan Paul.

Wolf, A. 1975. Obituary of Maurice Freedman. *Royal Anthropological Institute News* 10: 11–12.

INDEX

9 781845 456993